EESES

Electric Energy Systems and Engineering Series

Editors: J. G. Kassakian · D. H. Naunin

E. Handschin · A. Petroianu

Energy Management Systems

Operation and Control of Electric Energy Transmission Systems

With 43 Figures

Springer-Verlag
Berlin Heidelberg NewYork
London Paris Tokyo
Hong Kong Barcelona Budapest

Prof. Dr.-Ing. Edmund Handschin
Universität Dortmund, Lehrstuhl f. elektrische Energieversorgung
Emil-Figge-Straße 70, W-4600 Dortmund 50, Germany

Prof. Dr.-Ing. Alexander Petroianu
University of Cape Town, Dept. of Electrical & Electronic Engineering
Private Bag Rondebusch
7700, Republic of South Africa

Series Editors:

Prof. J. G. Kassakian
Massachusetts Institute of Technology,
77 Massachusetts Ave., Cambridge, MA 021329, USA

Prof. D. H. Naunin
Institut für Elektronik, Technische Universität Berlin
Einsteinufer 19, W-1000 Berlin 10, FRG

ISBN-13:978-3-642-84043-2 e-ISBN-13:978-3-642-84041-8
DOI: 10.1007/978-3-642-84041-8

Library of Congress Cataloging-in-Publication Data
Handschin, Edmund. Energy management systems :
operation and control of electric energy transmission systems / E. Handschin, A. Petroianu.
p. cm. -- (Electrical energy systems and engineering series)
ISBN-13:978-3-642-84043-2

1. Electric power systems--Management. 2. Electric power transmission.
3. Electric power distribution.
I. Petroianu, A. (Alexander). II. Title. III. Series.
TK1005.H295 1991 91-17377 621.31--dc20

Softcover reprint of the hardcover 1st edition 1991

Typesetting: Camera ready by authors

61/3020-543210 Printed on acid-free paper.

Introduction to the Electric Energy Systems and Engineering Series

Concerns for the continued supply and efficient use of energy have recently become important forces shaping our lives. Because of the influence which energy issues have on the economy, international relations, national security, and individual well-being, it is necessary that there exists a reliable, available and accurate source of information on energy in the broadest sense. Since a major form of energy is electrical, this new book series titled *Electric Energy Systems and Engineering* has been launched to provide such an information base in this important area.

The series coverage will include the following areas and their interaction and coordination: generation, transmission, distribution, conversion, storage, utilization, economics.

Although the series is to include introductory and background volumes, special emphasis will be placed on: new technologies, new adaptations of old technologies, materials and components, measurement techniques, control – including the application of microprocessors in control systems, analysis and planning methodologies, simulation, relationship to, and interaction with, other disciplines.

The aim of this series is to provide a comprehensive source of information for the developer, planner, or user of electrical energy. It will also serve as a visible and accessible forum for the publication of selected research results and monographs of timely interest. The series is expected to contain introductory level material of a tutorial nature, as well as advanced texts and references for graduate students, engineers and scientists.

The editors hope that this series will fill a gap and find interested readers.

John G. Kassakian · Dietrich H. Naunin

Foreword

The network control is a young discipline and yet there are already a considerable number of textbooks published on this topic.

The present lecture notes represent a unified approach to Energy Management Systems (EMS) from the hardware, software and human operator viewpoint. The book provides a framework within which EMS are to be realised, considering the state-of-the-art and the future developments taking place in this multidisciplinary field. A great emphasize has been put onto the systematic description of the different operational planning aspects. The content of this book is restricted to network control of electrical transmission systems. The interesting and far reaching problems of distribution network control differ very much from transmission systems and are not considered here.

In order to illustrate the specific operational aspects and requirements of some typical utilities, three different cases have been chosen from North America, Europe and from a developing country. The justification and the performance of a network control centre are very difficult to evaluate in terms of an economic cost-benefit; therefore, in the book, an attempt is made to answer these questions from a technical point of view.

The topic of project management is of utmost importance because an energy management systems brings together two very different engineering fields with quite different life cycles: power engineering experiencing a long life time in generation, transmission and distribution equipment and computer engineering characterised by short life time cycle and fast technology changes. The book examines some salient features of the project management activity

This book may serve as lecture notes for a graduate course in energy management systems as well as a concise introduction to engineers who enter the field of power dispatch control. Last but not least it may serve as a guide for practitioners from the electrical utilities, manufacturers, vendors and consultants involved in the different aspects of the design, development, implementation and operation of a network control centre.

A great effort has been put into the edition of a glossary which contains the most important concepts used in the field of energy management systems. This part of the book represents a contribution to the effort trying to unify the terminology in this new and fast growing engineering field.

Contents

1 Introduction

1.1 Electric energy systems

The high technical standard of electric energy systems is based on two requirements:

- Advanced technology of the components comprising all parts of the power system (primary technique)
- Implementation of high level information processing methods to implement energy management systems (secondary technique).

The optimal combination and coordination between these two fields is a major engineering challenge for the implementation of a modern electric energy supply system.

Fig. 1.1-1 shows schematically the application of information processing in the area of

- Power plant control
- Power system control
- Substation control
- Demand side load control.

While the electric energy flows from the generation through the transmission system to the loads, the information flows in both directions: from the components to the control centre (telemeasurement) and from the control centre to the power plants and substations (telecontrol). The interconnection between energy and information is given by the technical and economical requirements of electric energy systems.

Compared to power engineering, advanced digital information processing is a relatively young engineering discipline. However, modern power systems need to be operated with a well-developed information technology; but even the most advanced information technology cannot improve the operation of a poorly planned and/or implemented power system.

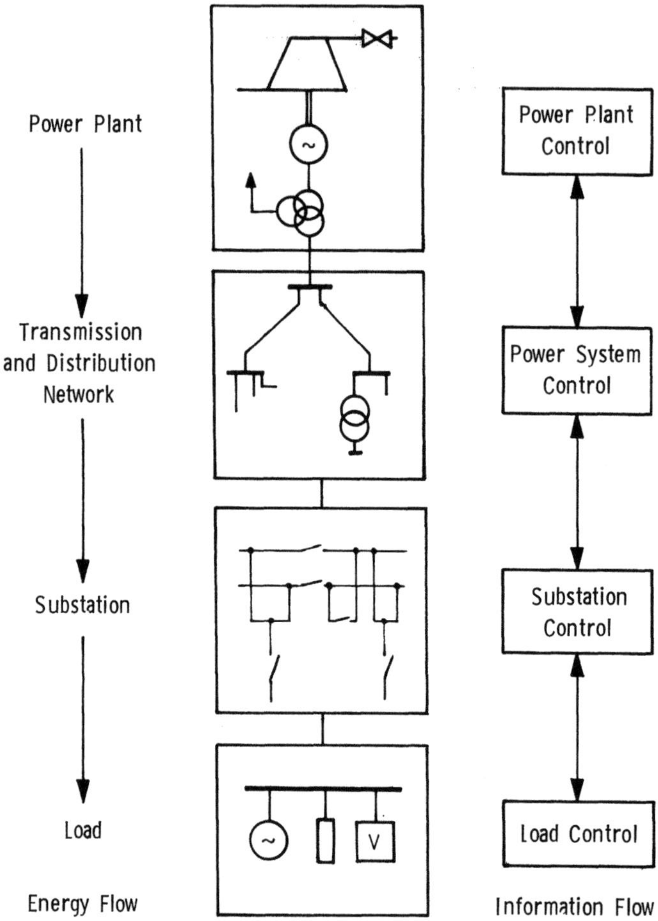

Fig. 1.1-1: Energy and information system

The hierarchical structure of the different power engineering control systems is shown in Fig. 1.1-2. It consists of the decentralised tasks of power plant control, substation control and load control, on one side, and the centralised power system control on the other. In both areas the main classes of

- Information processing
- Monitoring, protection, command and control

are evident. Without digital data transmission and processing systems the realisation of these control systems is inconceivable.

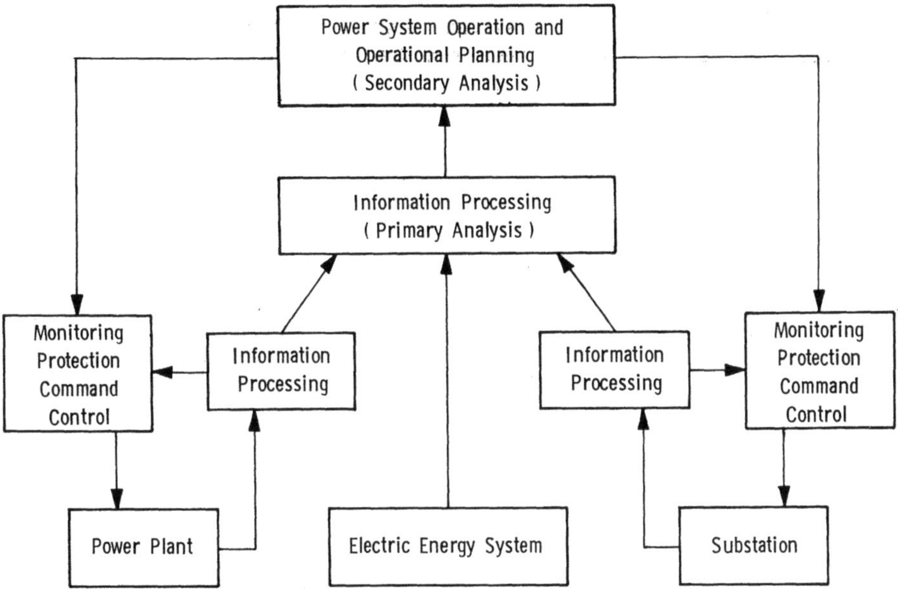

Fig. 1.1-2: Hierarchical structure of power system control

A modern energy management system (EMS) is shown in Fig. 1.1-3. The power system measurements are transmitted via a digital information system to the central data base. The primary analysis covers the following tasks:

- Programs for message switching
- Control of the information display systems
- Recording of status and events
- Information processing for telecommand and control.

The advanced decision support and optimisation tools in secondary analysis include:

- State estimation
- Optimal power flow
- Network security analysis
- Load prediction

- Long-, medium- and short-term optimisation including unit commitment and economic dispatch.

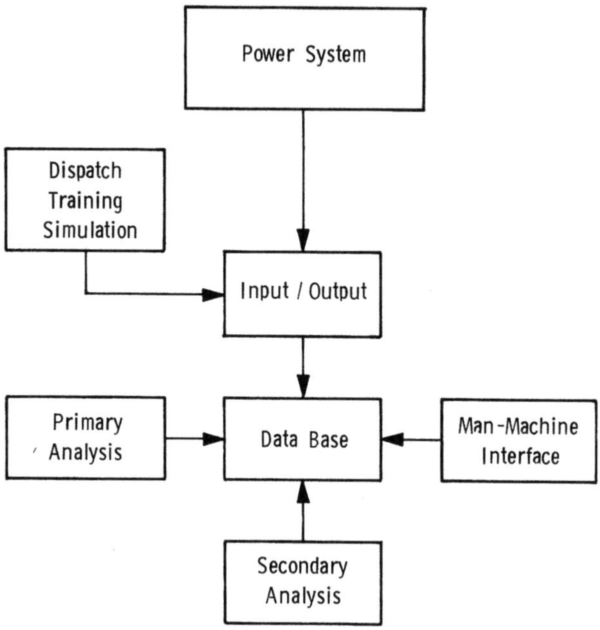

Fig. 1.1-3: Energy management system with the primary analysis for monitoring and the secondary analysis for system operation and operational planning

1.2 Power system engineering

Power system engineering is not solely concerned with individual components of a power system such as power plants, power lines, transformers etc. but within the operational aspects of the entire power system functioning with quality, economy and security related constraints. In general, it seeks solutions defined by the following four criteria:

- Economy of operation
- Security of supply
- Quality with respect to frequency and voltage level
- Environmental compatibility.

System engineering describes the operational states of a power system (see Fig. 1.2-1) according to the four classes shown in Table 1.2-1.

Table 1.2-1: Operational states of a power system

State	Energy Equilibrium	Technical and Economical Constraints	(n-1) Principle
Normal	satisfied	satisfied	satisfied
Alert	satisfied	satisfied	not satisfied
Disrupted	satisfied	not satisfied	not satisfied
Network splitting	not satisfied	not satisfied	not satisfied
Restoration	not satisfied	satisfied	not satisfied

The (n-1) principle implies that the loss of one power system element does not lead to a power system disruption.

For transition from alert to normal state, preventive actions by the control engineer are needed to avoid disruption as the result of a possible fault in the power system. If the power system is disrupted, corrective actions must be taken because important technical and/or economical constraints are violated.

To take correct action implies a very complex decision-making process, often resulting in an unacceptable time pressure on the human operator. Hence dispatch training simulators form an important part of a modern energy management system (see Fig. 1.1-3)

The basic features of a modern EMS may be described as follows:

- Monitoring the system state by means of analog and digital information
- Interaction with the power system in a preventive or corrective manner
- Regulation of frequency and voltage
- Reliable and complete system information despite wrong or missing data
- Evaluation of operational risks for the present or future state

- Actions to improve power system security
- Economic power system operation
- Optimal primary energy usage observing environmental aspects.

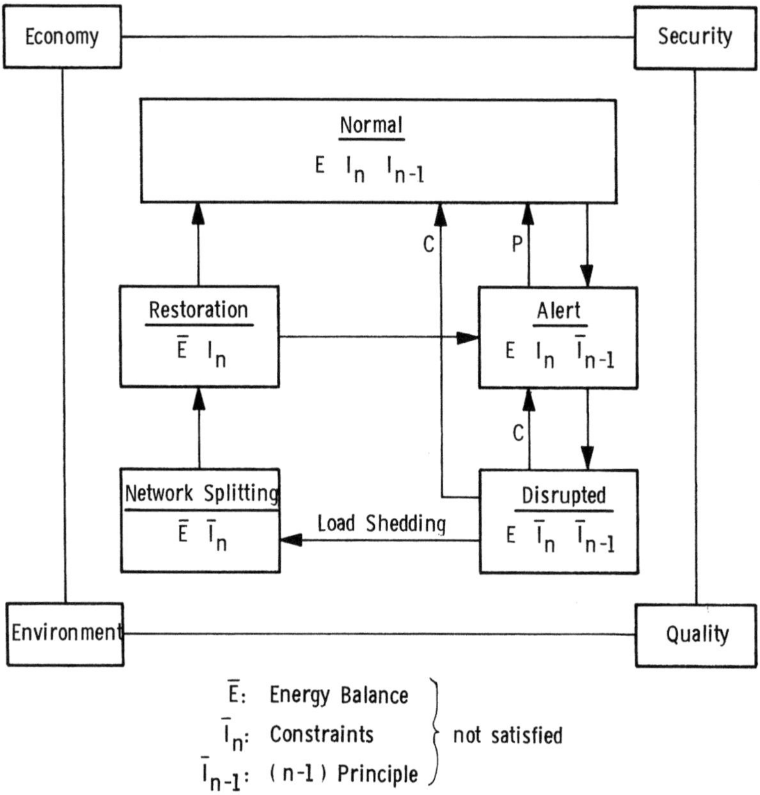

Fig. 1.2-1: Power system states
P : preventive measures
C : corrective measures

The necessity for EMS arises from many factors affecting power system operation:

- Future development of electric energy demand in industrial and developing countries
- Availability of primary energy resources for power generation
- Rapidly changing economical and political constraints when implementing expansions of electric energy systems

- Functioning of the power system within operational limits because of limited availability of generation and/or transmission capacity due to restrictions in the necessary system expansion
- Strong impact of new technology based on microelectronics for a more reliable and economical operation as well as rational use of electricity
- Increased use of information technology to handle the complex decision-making process.

1.3 Evolution of power system control technology

The most important significant events in the evolution of power system control technology are represented in Fig. 1.3-1. Until about 1940 the dispatcher of the power system was located in a power station or large substation. The advent of power system control technology evolved through local monitoring and control, the use of a static mimic board and the telephone for commands to the field operators. The evolution of data acquisition and remote control starts with analogue technique. Automatic generation, interchange and frequency control also uses the analogue technique. During the period 1950-1970 digital computers were extensively used for off-line power system planning studies. The New York blackout (1965) forced the power utilities to reconsider on-line reliability problems, the most important consequence being the accelerated introduction of SCADA/EMS. Following the introduction of computers into the power plants, process computers and the graphical displays were also implemented in power dispatch control centres.

The period after 1970 was very fruitful in the development of state estimation and optimal power flow theory. The second great blackout in the USA in 1977 reinforced the importance of network security assessment. Other blackouts and incidents in Europe highlighted such aspects as dispatcher training simulators, corrective actions in emergency situations or the importance of voltage stability. The Three Mile Island incident emphasised the importance of human engineering in displaying the information to the operator. SCADA/EMS are also penetrating the medium-voltage networks. However, the tasks of distribution automatisation are different from the specific functions of the SCADA/EMS in high-voltage power transmission systems. There is a growing awareness of the need to unify dispatch of supply and demand (economic dispatch and load management).

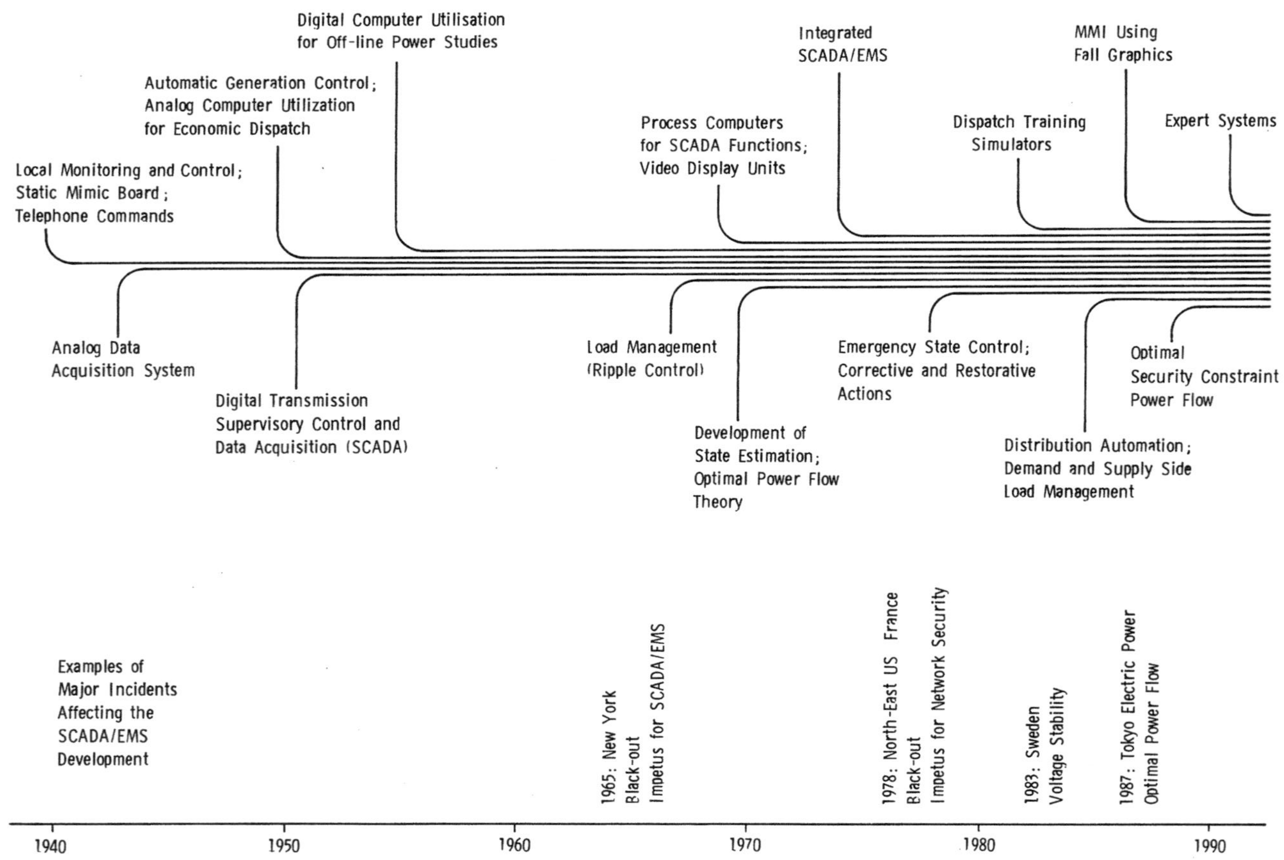

Fig. 1.3-1: Significant events in the evolution of power system control technology

The necessity to operate the power system closer to technical and economic limits accelerated the implementation of optimal power flow considering security constraints present in EMS. The different objective functions for the optimisation include:

- Real power losses
- Equal voltage profile
- Prespecified short-circuit capacity.

Powerful computer hardware permits the realisation of optimisation procedures in real-time. However, efficient handling of sophisticated EMS requires highly skilled and trained operators.

The use of knowledge-based systems into SCADA/EMS constitutes the latest developments in power dispatch control technology. It offers a new form of implementing algorithmic procedures and heuristic experience. Knowledge-based information processing supports the complex decision-making process of the power system operator.

1.4 Control centre justification

1.4.1 Associated effort

A major trend throughout the electric utility industry is the replacement of conventional dispatcher's offices by modern control centres. Considerable progress has been made in developing dispatch control technology. By 1990, more than 300 control centres have been implemented or planned. This number is still too small compared with the number of world-wide existing utilities. Cost, of course, is a major deterrent but not the only one. Table 1.4-1 shows some of the difficulties and efforts associated with the introduction of a new control centre. Although it is already possible to show a positive cost/benefit ratio in some SCADA/EMS applications, in most cases there are not yet enough statistical data for a comprehensive cost/benefit analysis. Hence a sound methodology for a control centre justification is still missing. Some specialists support the opinion that a quantitative justification is so complex that it is becoming almost impossible to accurately measure the value of control centres.

Table 1.4-1: Quantitative Analysis of an EMS

Investment

A new control centre represents an investment of between US $2 and 40 million. The full graphics add about $ 0.5 to 2 million. Cost estimation is difficult due to the lack of standardisation. Each new SCADA/EMS project signifies changes in approximately 25% of the SCADA environment and 100% of the data base.

Size

A medium SCADA system contains about 50 remote terminal units (RTU).

Complexity

A medium SCADA/EMS software system contains over 1 million lines of codes.

Development Effort

A network security package represents more than 50 man-years development effort. An optimal power flow package represents about 15 man-years development. The power application software package requires approximately a million lines of FORTRAN code. From the supplier's contribution, over 200-300 programmers are needed for 3-5 years to develop a new SCADA/EMS system.

Database

A medium-size data base contains up to 10^8 data items. A power application package requires 1-10 MB of data memory. For a data base containing 350 000 data items approximately 13 000 man-hours are necessary for compilation and debugging. Updating requires continuous work for 1-2 specialists.

Man-Machine Communication

The displays are user specific and highly customised. For a medium-size network approximately 300-500 displays are designed, coded, tested and up-dated. A large utility could require 5000 different displays. The preparation of displays requires approximately 3-5 hours per display; approximately 0.5 man-day is necessary per substation diagram design and implementation. In addition, displays which are laid out on paper look considerably different from those on a CRT screen. Consequently, it may be necessary to redesign and rebuild the displays several times.

Time

Between ordering the EMS and the fully operational state, the time taken could be between 2 and 10 years.

Staff

Supplier contribution: 5-10 specialists
User contribution : 2- 5 specialists.

Documentation

5-10 meters of shelving for project documentation.

Performances

Difficult to estimate. The only true benchmarks are measurements of the complete working system.

Maintenance

10-15 years.

1.4.2 Factors justifying a new control centre

The factors that contribute to the need for a modern control centre can be divided into the following categories:

- The growing complexity of operating problems
- Enhancement of system economy
- Enhancement of system reliability

With the increase of:

- Generating plants
- Utilisation of EHV transmission lines
- Increase in number and the importance of interconnecting tie-lines

power system operation is becoming a more complex task.

The constraints imposed on the utility's operating environment in recent years for various security, economic, environmental and regulatory reasons have made the tasks of the system operator increasingly more difficult. Power systems are operating closer to their design capabilities and security margins. This, coupled with continuously escalating fuel costs, made it mandatory for utilities to operate the system as optimally as possible. In this operating environment, more complex decision-making processes are required. Operating personnel need more timely and meaningful information on which to base their numerous and important decisions. If the utility has just suffered a major outage, the management will be under pressure to prevent any future blackout and any decision for a modern control centre to be adopted will be fastened.

It is worth noting the analysis of specialists in Finland; according to their experience, the costs of one hour's national wide outage are about the same, as far as the national economy is concerned, as the total costs of the control centre (i.e. approximately US $25 million). Not all the specialists agree with such a simple justification; unfortunately there is no standard or simple methodology for evaluating the economic worth or the cost-effectiveness of a new control centre. The exact benefits have not been fully evaluated, but it should be mentioned that the cost of a new control centre represents only a small amount of the utility's total equipment costs. Building a modern control centre is often compared to buying relatively inexpensive insurance for the large capital investment in the utility's generation-transmission system, the security-oriented functions being regarded as an insurance against interruptions and blackouts in the power systems.

An American study attempted to measure the benefits of a SCADA/EMS control centre by evaluating the benefits from both operating (e.g. production costs) and planning aspects (capital costs). For different systems ranging between 3600 MW and 15 300 MW the daily savings may be calculated through economic dispatch facility using a dynamic programming method. This comparison shows savings in the order of a half percent in fuel costs. The study also examined the economic effects of errors in the load forecast with the conclusion that very accurate load forecasts are not required. An important contribution of the study was the measurement of the value of a security analysis program within the EMS. Today the planning process looks to add new elements in order to maintain a level of security against such contingencies assuming present-day operation. The study examined the cost savings an EMS with security dispatch could produce if the planning process recognised the capabilities of that system. This produced the capital cost differences attributed to security dispatch functions. The study also examined the effects of security dispatch on production costs; this was measured by seeing if it could respond to contingencies more economically and hence simulate operation. Finally, the study examined how well an operator would monitor the power system "with" and "without" security dispatch. The security dispatch within the EMS improve reliability (i.e. it prevents load outages or service interruptions). The cost of those outages was used as a measure of the value of security functions.

The approach taken in the study is to accomplish planning for transmission expansion so as to maintain the same level of reliability with and without an EMS. The planning of transmission expansion for a period of ten years, assuming a manual or operator-based security monitoring, leads to a total cost of about US $172 million. The planning of transmission expansion for the same period, assuming an EMS with security functions, involves a total cost of US $15 million. The study concludes that the present improvement value which the security dispatch function produced is approximately US $150 million. Table 1.4-2 summarises the effects of security dispatch on production cost for some outages.

Table 1.4-2: Operating cost effects

Case	Costs	
Outage A	Base operating cost........	100%
	Operating cost using manual/ operator redispatch...........	107.1%
	Operating cost using EMS security functions............	101.3%
Outage B	Base operating cost...........	100%
	Operating cost using manual/ operator dispatch............	107.9%
	Operator cost using EMS security functions...........	103.1%

The cost analysis of security dispatch functions with respect to production costs shows that in some cases the increase in cost due to operator redispatch is over 7%, as opposed to much smaller increases when the corrective redispatch is performed by the security constrained economic dispatch program. Other benefits achieved by the SCADA/EMS functions are shown in Table 1.4.-3.

Table 1.4-3: Benefits achieved by EMS

SCADA

Required to support all other functions. Decreased labour costs, or more effective use of labour, especially in distribution. Improves the customer supply service. Can prevent equipment damage. Shows overloading. Improves security by giving operator timely information. The operator's response in dealing with supply interruption is much faster.

Automatic Generation Control

One US utility estimated that a digital control system produced savings of the order of US $30 000/month compared with the previously used analog system.

Interchange Scheduling

More than 10% of transacted energy values saved.

Economic Dispatch

For a utility with an annual $1 billion fuel cost, a realistic assumption is a one-tenth of 1% of the fuel costs, i.e. $1 million saving/year.

Unit Commitment

Sample calculation has shown a fuel-cost reduction of 0.5-2 %.

Load Forecasting

A 5% forecast error in peak load may be enough to cause the scheduling of an extra generating unit. A load forecasting program could prevent the starting of this extra generating unit.

Transmission Loss Reduction

A utility serving an average load of 5000 MW may be losing 50 - 150 MW. This translates up to US $140 000/day at a typical $40/MWh cost of produced energy. In developing countries, only 1% better efficiency in a power network will mean that, for instance, 50 MW of costly diesel generating units can be shut down.

Reactive Power and Voltage Control

For a utility with an average load of 5000 MW whose losses are 50 - 150 MW, optimising reactive power flows can reduce losses by about 3-5% or 1.5 to 5 MW. Saving only 1 MWh per hour is worth some $350 000 each year if generating power costs are $40/MWh.

State Estimation

The Swedish State Board implemented state estimation (SE) in 1979. From 1979 to 1981, on the basis of SE, about 10% of the collected measurements were corrected. Today, during normal operation a faulty instrumentation (wrong data) is detected, with the help of SE, twice a month. For the Belgian control centre the cost of developing SE has been found to be equivalent to one kilometre of a double 115 kV line.

Network Security Assessment

EdF estimates that, at the level of the year 1981, network security analysis prevented the starting up of at least one 600 MW fuel unit for approximately 4500 hours. The resulting gain is much greater than the development and the annual operating cost of the SCADA/EMS system.

Optimal Power Flow

Ontario Hydro conducted a detailed study to assess the possible saving in transmission losses if an on-line optimal power flow program is used. The results show that up to 3.7% reduction in total system losses could be achieved, resulting in a $2.5 million annual saving.

1.4.3 Conclusions

There is widespread consent between utilities that automation and the introduction of SCADA/EMS based modern control centres is the way of the future for improving the service level. However, for the the benefits to be recognised, there may be many intangible factors which are not quantifiable or only quantifiable with difficulty. Today there is no generally accepted methodology for evaluating the cost-effectiveness of a new control centre. Table 1.4-4 summarises the advantages and disadvantages associated with the introduction of a new control centre.

Table 1.4-4: Evaluations for the introduction of a SCADA/EMS based control centre

ADVANTAGES	DISADVANTAGES
- provides more information	- creates more work
- is more effective	- increases the amount of information to be monitored
- easier to evaluate the power system state	- requires more complex tasks to be fulfilled
- allows faster decision-making	- puts higher requirements on the operator's qualification
- improves the response time to system faults	- creates more stress under emergency conditions and poor control centre system design
- system problems may be avoided through use of prophylactic functions	
- better record keeping	
- more flexibility to system changes	

2 System engineering aspects of power system operation

2.1 Classification

The vast amount of different power systems with such a variety of structures, design, geographical features, voltage levels, load patterns, equipment characteristics, sizes, topological configurations, organisations, policies, operating rules, differences in personnel qualifications, etc., cannot be mapped, either quantitatively or qualitatively, into a single control centre design and into a unique SCADA/EMS functional structure.

The diversity and the plurality of power systems involve a diversity of network control centre functions. However, the question is, how to cope with such a diversity of network control centre functions. The answer is by categorising, classifying, and breaking these functions into a number of frames and coordinates. In order to achieve a systematic presentation a classification of the SCADA/EMS functions will be introduced in the sequel. As this is a very difficult undertaking due to many possible criteria and different and conflicting perspectives or viewpoints, it is not possible to determine a unique classification criterion. A "decoupled" approach has been followed by examining the SCADA/EMS functions in different "subspaces" or "frames" of coordinates. Table 2.1-1 summarises this approach.

Table 2.1-1: Classification of the SCADA/EMS functions

Classification or criterion related to	Frame or Attributes
Time dependency	Pre-dispatch; dispatch; post-dispatch. Real-time; extended real-time. Preventive; corrective. Post-mortem/post-factum analysis.
Type of network	Transmission; subtransmission; distribution.

Mode	Operation/dispatch mode; Operational planning Simulation or engineering mode; Training mode
Main tasks of the operation and operational planning	Power management; switching management; Supply and demand side management; Management of the network
Level of the dispatching hierarchy	Corresponds to the various voltage levels and different territorial structures
Closeness of the SCADA/EMS functions to the process	On-line; closed-loop off-line
Structure of the power dispatch staff	Operators; Supporting staff
Priority of the functions	Primary analysis; Secondary analysis
Utility's control policy	Centralised; Decentralised; Federalistic
Power system operating state	Normal; alert; disturbed; network splitting; restoration
Functional structure	Security functions; economy functions; Functions related to the quality of supply; Personnel training functions

2.2 Time decomposition

The information required and the specific operator's activities are heavily influenced by the time period which accordingly varies with the amount and the requirements of the SCADA/EMS functions. The conditions of the power system can conveniently be divided into three different periods:

- Pre-dispatch.......period of time before the real-time operator's activity
- Dispatch...........period of time during the real-time operator's activity
- Post-dispatch......period of time after the real-time operator's activity.

The general operator's activities and SCADA/EMS functions associated with each of these periods are discussed in the following sections.

2.2.1 Pre-dispatch

The most important activities and functions are:

- Short-term load forecasting
- Generator scheduling
- Interchange planning
- Maintenance activities planning
- Reactive power and voltage-level planning.

In the pre-dispatch phase, the operational planning develops, provides and maintains a short-term plan to meet time-related needs for the following hour and for subsequent periods. The pre-dispatch functions and tasks support the real-time decision-making process; therefore they are also called predictive, prophylactic, preventive and "look ahead" functions.

2.2.2 Dispatch

The most important activities and functions are:

- Data acquisition and storage
- Power system monitoring
- Power system frequency and tie-line control
- Interchange monitoring
- Economic dispatch control
- Power dispatch (supply and demand side management)
- Switching activities

- Power system disturbance monitoring; power redistribution; remedial actions; corrective switching; load shedding
- Protection related activities.

In the dispatch phase, the operator implements the plans developed in the pre-dispatch period and manages the resources to provide electricity and meet the power system needs at the present time.

The dispatch functions run in real-time in order to cope with the power system and the operator's time requirements. The dispatch functions support the operator in his decision-making process.

2.2.3 Post-dispatch

The most important activities and functions are:

- Operator activity logging
- Energy accounting
- After-the-fact problem analysis
- Report generation.

The post-dispatch functions and tasks result from the real-time decision-making process of the dispatcher; their conclusions are used for improving (enhancing) the dispatch and the pre-dispatch functions.

The post-dispatch functions represent a post-mortem or after-the-fact analysis of the process. The pre-dispatch and the post-dispatch modes possess the following salient features:

- They work in an extended real-time environment
- They are not so close to the on-line process
- They have a predictive (prophylactic), or a post-mortem (post-factum) analysis, and study (engineering) character.

There is a strong correlation between the three modes. There may be some time overlap between the three periods. Fig. 2.2-1 shows a graphic illustration of the correlation and the functions provided in the pre-dispatch, dispatch and post-dispatch period.

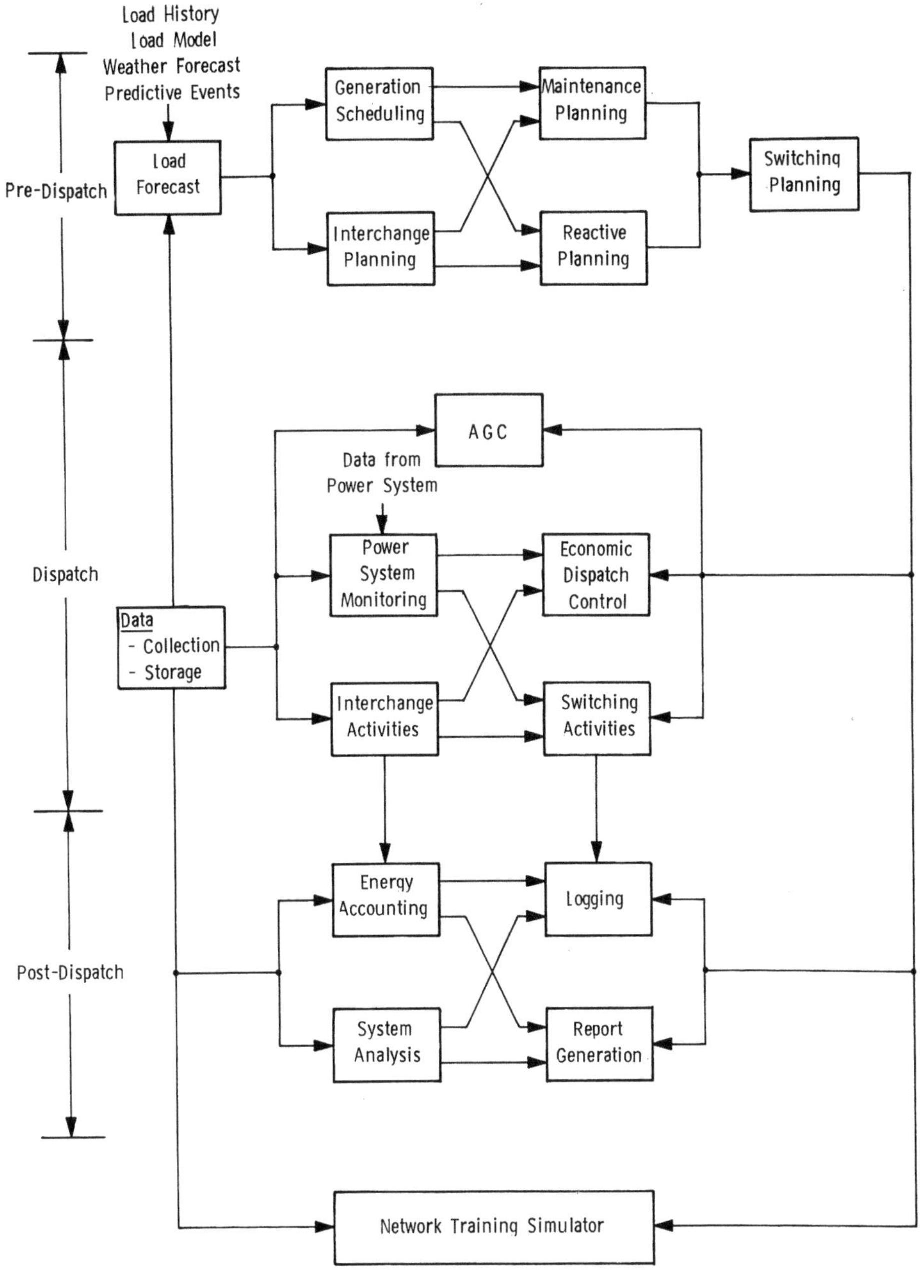

Fig.: 2.2-1: Time decomposition of the power system operation

2.3 Network level decomposition

The functions and tasks of a SCADA/EMS system can be classified according to the network level. A power system may be broken down into the three levels: transmission, sub-transmission, and distribution.

2.3.1 Transmission

The transmission network covers all the territory served by the utility and performs long-distance energy transfers; the main part of generation is connected directly to this network. The transmission network is meshed. The voltage level is 220 kV, 380 kV and higher.

2.3.2 Sub-transmission

The sub-transmission network covers a part of the territory connected to the transmission network and performs medium distance energy transfers; a minor part of generation is connected to this network. It is often operated partially meshed and in some cases radially. The voltage level lies between 20 kV and 220 kV.

2.3.3 Distribution

The distribution network covers the local transmission of energy on short distance; in general, only small generation units are connected to this level. The loads of the distribution network can be classified into the following types of consumption: regional, municipal, industrial, rural, and railway.

The distribution network normally has a radial structure. The voltage level is between 0.2 kV and 110 kV.

2.3.4 General remarks

It must be pointed out that the requirements, responsibilities, objectives, policies and procedures of the transmission, sub-transmission and distribution networks are quite different. Therefore SCADA/EMS have quite different structures, features and functional architectures.

There is a long-term trend to down-grading the network (from transmission to sub-transmission and from sub-transmission to distribution) in conformity to the load increase and density; location of generation plants and load centres; change of the type of production sources; and other factors. There is also a long-term trend to up-grading the control centre functions, i.e. some functions exercised at the transmission level will be performed/implemented at sub-transmission and even at the distribution level. On the other hand specific functions for the distribution level are being introduced at the transmission level (e.g. load management).

2.4 Mode decomposition

There are two modes: operation and operational planning. Operation corresponds to the real-time period; the operator or the dispatcher takes decisions and acts in the real-time frame. Operational planning corresponds to both the pre- and post-dispatch time frame.

2.4.1 Operation

This mode is related to decision-making and actions in real-time concerning such tasks as power control and commands, maintenance of quality of energy (frequency, voltage) supervision of the system state, switching control or commands, security evaluation, load management, decision-making as far as safety of people is concerned, and remedial activities.

2.4.2 Operational planning

This mode is not related to the real-time frame but to long-, medium- and short-term scheduling and planning for the real-time operation mode. It supports all the tasks and decision-making processes taking place in real-time. If longer periods are considered, one progresses from operational planning to expansion planning tasks.

2.5 Operation state decomposition

The broad definitions applied to the operating states of the power system as used in the literature are shown in Fig. 1.2-1.

In the normal state, the security and economic goals are being met within established constraints. In the degraded state, a series of three levels depending on the severity of the non-normal power system problems may be defined as:

- Alert
- Disrupted
- Network splitting (collapse).

Information required and the actions taken by the operator vary according to the state of the system. During the normal operating state the operator's attention will concentrate on preventive actions. During system degradation the operator's attention will be focused on such functions as disruption monitoring, corrective control, remedial actions, corrective switching, power redistribution, load shedding, and on information that is critical for this aspect of the power system.

2.6 Activity decomposition

As power systems become larger, there is a need for the power system operators to specialise in one of the following two main management tasks:

- Power management
- Network management.

2.6.1 Power management

These activities encompass all the functions and tasks connected with supply (generation) and demand (consumption) side management such as load forecasting, automatic generation control, economic dispatch control, interchange scheduling and accounting, unit commitment, transmission losses minimisation, power wheeling, power brokerage, load management and ripple control, load shedding, voltage and reactive power scheduling, monitoring and control.

2.6.2 Network management

These activities encompass all the functions and tasks connected to switching such as switching requests, switching decisions, switching executions. The generic term "switching" covers all the actions changing the network topology and all the decisions influencing the safety of the people involved in the maintenance of the elements in the network. The switching operator is in charge of the overall system security and integrity in the normal state and for all the corrective switching and actions in emergency and restorative situations. The following network security functions are closely interwoven within the network management: network configuration, state estimation, limit check, dynamic network colouring, on-line load flow, contingency screening, contingency evaluation, short circuit analysis, network equivalencing, stability monitoring, ampacity monitoring.

As shown in Fig. 2.6-1, coordination of the power management prevails at a higher decision level, while at a lower decision level execution of

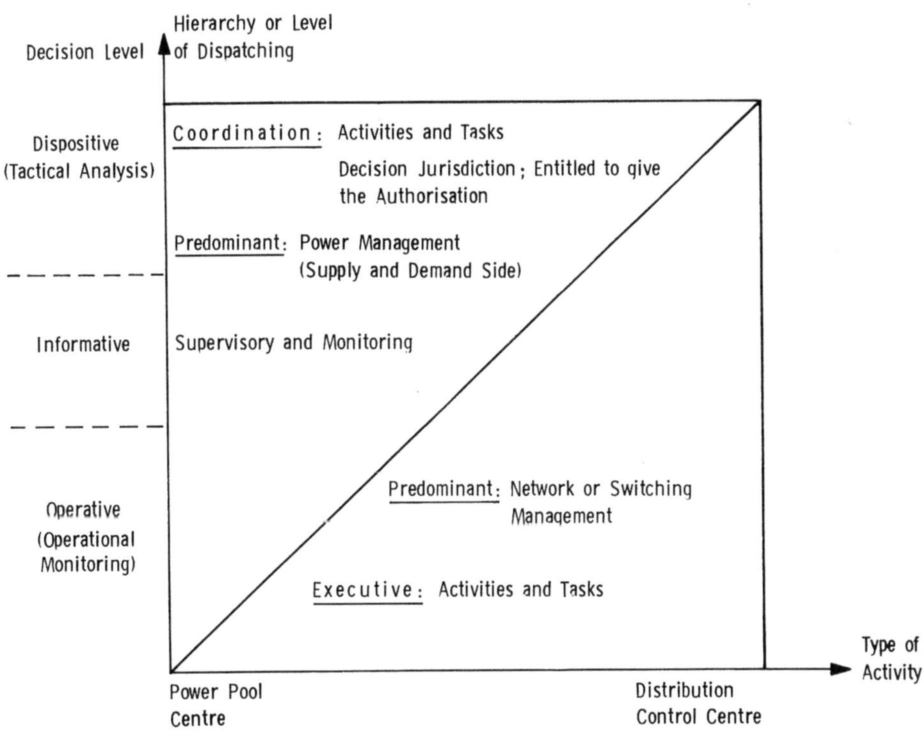

Fig. 2.6-1: Activity decomposition

the switching management occurs. However, this is a specific utility's own policy. There are systems where the switching coordination and execution are at the highest dispatching level.

2.7 Control decomposition

Decomposition, according to the utility's control policy, involves two types of control, namely, centralised and decentralised.

2.7.1 Centralised control

When decisions and executions are managed at the same level, control is centralised.

2.7.2 Decentralised control

When execution is on a lower level than the decision level, control is decentralised. Here again two types have to be distinguished:

- Decentralised or hierarchical controlled power systems with semi-independent area control.
- Decentralised with independent area control.

In a decentralised controlled power system with semi-independent area control, the system control centre performs a coordinating role (e.g. a large national utility, a power pool).

In a decentralised system control with independent areas, each area is responsible for the respective transmission network. The system control centre (if any) performs a loose coordinating role (e.g.. a multinational power system, a super-pool, etc.)

2.7.3 Centralised versus decentralised policy

Control centres were mainly designed for normal or alert states, allowing operators to optimise the transmission and generation system, thereby reducing the overall operating costs and increasing the security of the system. Practice has shown that the control centre design was not generally adequate for emergency periods, where the high rate of data flow could sometimes lead to poor computer response time and operator overload. Excessive centralisation is one of the reasons for this inadequacy.

Functions of the emergency type such as protections, load shedding, network splitting must be decentralised. This follows the trend to decentralise what can be realised locally. An electrical power system is inherently integrated, therefore, decentralised and centralised functions need to be coordinated. There is a tendency to centralise such functions that cope with the natural coupling of the system elements in a normal state (economic dispatch, automatic generation control, security assessment).

Otherwise, the emergency and restoration control functions are mainly decentralised due to stress, overload of operators, speed requirements, etc. Coordination of the decentralised functions is one of the trends in power dispatch centre technology. It is of the utmost importance to re-adapt the local automatic device set points in real-time to the states and constraints operating within the system.

2.8 User oriented decomposition

In a control centre there are two kinds of users or personnel:

- Power system operators or dispatchers
- Supporting staff or operational planning engineers.

Real-time and on-line open-loop functions such as dynamic network colouring or state estimation are used by the operators. The extended real-time functions can be used by both the operators and the operational planning engineers.

2.9 Analysis decomposition

2.9.1 Primary analysis functions

Functions that perform the primary analysis are those designed to collect and display data, check the limits, and log and store data. Primary analysis is a connotation for SCADA functions.

2.9.2 Secondary analysis functions

These are functions which use complex mathematical models, require coherent data sets and use sophisticated numerical methods. Secondary analysis is a connotation for EMS functions.

2.10 Control flow decomposition

The examination of the data and the control flow between the power system and the control system leads to the decompositions shown in Fig. 2.10-1.

The "0" data domain vicinity serves as input for all critical real-time and on-line closed loop functions: power system frequency and tie-line control, automatic generation control, load shedding, emergency control, dynamic network colouring (DNC), power system monitoring, limit check, alarms.

The "1" data domain vicinity serves as input for the real-time and on-line open-loop functions: network topology determination or network status analysis, state estimation, observability check, wrong data identification, network splitting, monitoring. This represents a schematic mapping of the process raw data into a time consistent data set and is used in the dispatch mode.

The "2" data domain vicinity represents the output of the "state estimation" function, the results of which form a secure, accurate and consistent data set.

The "3" data domain vicinity represents copies or modified copies of the "2" domain. These copies serve as input for the extended real-time functions. The whole range of the network security assessment and network economy utilise this data domain.

The "4" data domain vicinity serves as output for all the simulation, training and studying functions.

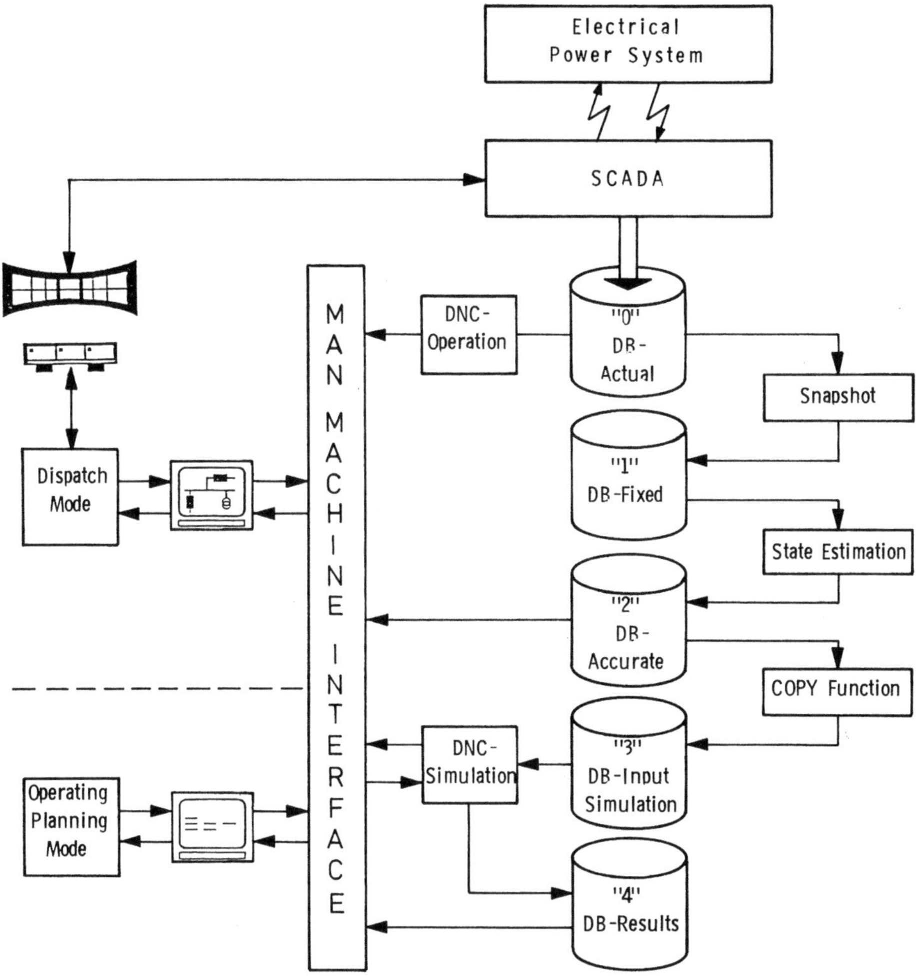

Fig. 2.10-1: Control flow decomposition

Fig. 2.10-2 represents the classification of the control functions into on-line and off-line tasks together with the real-time and extended real-time functions.

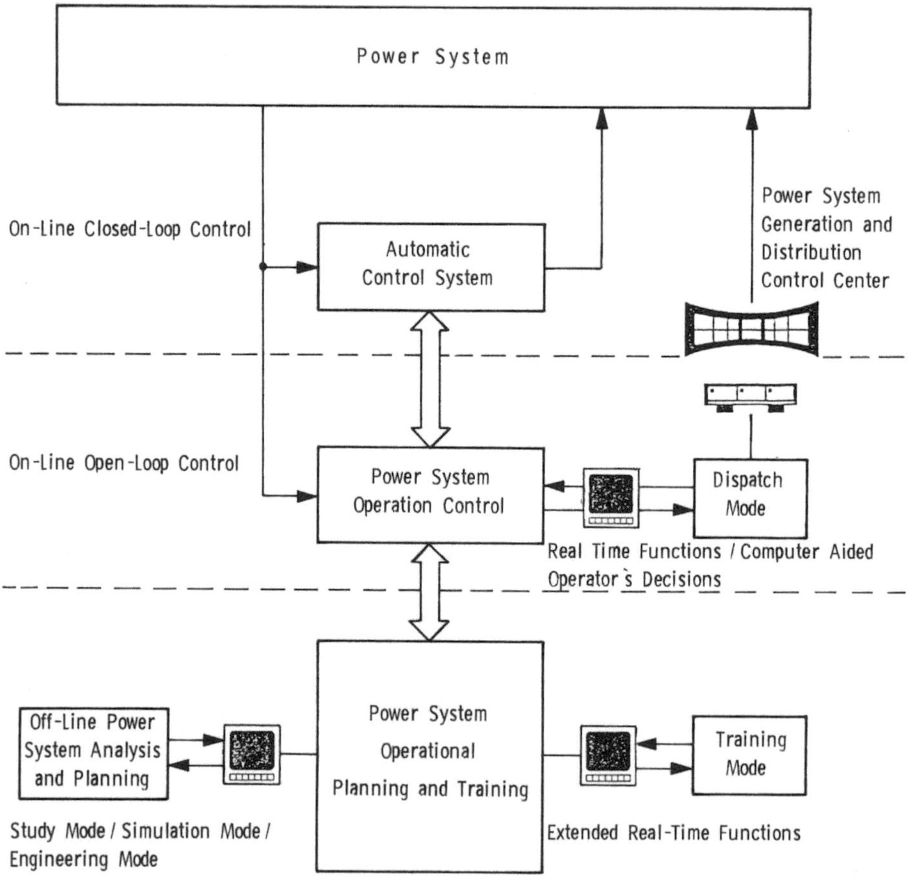

Fig. 2.10-2: Classification of the power system control tasks

3 Typical energy control centre functions

3.1 System monitoring and security

The concept of power system monitoring and security covers three aspects:

- Monitoring
- Assessment
- Control.

Security monitoring is based on real-time system measurements obtained from the SCADA system and filtered through state estimation and wrong data detection and identification. If all state and system variables are within the technical limits, the system is said to be in a normal state.

If it is in a normal state, the next step involves the question of whether the system is secure or insecure with respect to the next contingencies. If there is at least one contingency which may create an emergency situation the operator has to determine what preventive action should be taken to make the system secure.

Once the system is in an emergency state, the operator has to determine the corrective action to bring the system back into the normal state. If this corrective action is not initiated in time, the system will suffer severe disturbance with a possible blackout in a part of, or in the entire network; the system load then has to be restored through restorative control. Special procedures have usually to be executed during restorative actions. In most cases the system restoration may take a considerable time before it is brought back to a normal state again.

The major components of an EMS dealing with security assessment and control are shown in Fig. 3.1-1. The monitoring starts from the real-time measurements of analog quantities such as voltages, active and reactive powers and currents. The real-time binary information describes the topology of the network. It includes the status of breakers and switches as well as the tap settings of power transformers. The analogue and the binary information is collected at remote terminal units at the substations and transmitted to the control centre. The incoming data have to be preprocessed in order to check their plausibility and consistency.

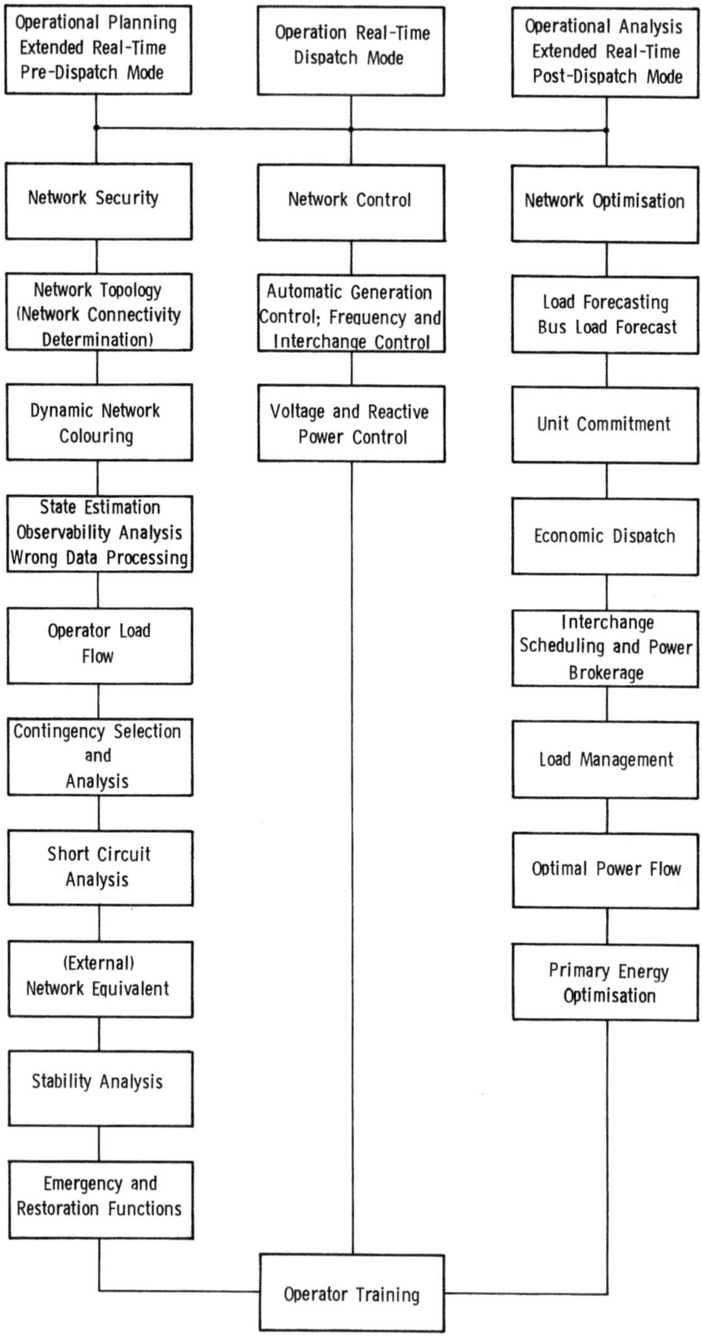

Fig. 3.1-1: Components of an EMS

The configurator processes the binary information in order to describe the network topology. This kind of information may be shown to the operator to explain the different parts of the power system. Furthermore, the configurator yields the mathematical model for the state estimator which processes the analogue measurements. The output of the estimator is the system state information based on the bus voltage magnitudes and the phase angles. All other system variables such as line currents, line power flows, active and reactive-power injections may be derived from the system state vector.

Closely related to state estimation is wrong data processing which aims to find out whether the measurement vector contains any incorrect data. These measurements have to be eliminated since they corrupt the reliable system information. At the same time the ability of the system state vector must be checked to establish whether state estimation can be performed for the entire power system or for which part of the system the state estimation may no longer be applicable since there are not enough measurements. In order to assess the power system state all important quantities have to be checked against pre-defined limits. These limits may be time-varying according to the specific properties described. The output of this activity indicates whether the system is in a normal, emergency or restorative state.

In order to discover whether the normal state is secure or not a set of contingencies has to be simulated. It is important to have an efficient method for the selection of critical contingencies. Contingency analysis currently involves steady-state load flow analysis; stability constraints are expressed in terms of the limits on line flows. It assesses the system's response to contingencies using the on-line load flow. The on-line load flow consists of its own power system model (including the data from the configurator and the state estimator) together with a real-time system representation of the external system.

3.2 System economy

Security control-based optimisation selects actions to prevent overload or to minimise the impact of contingencies. It is an optimisation problem based on an optimal power flow (OPF) analysis; hence it involves the minimisation of an objective function subject to specific operating constraints.

The purpose of an on-line OPF function is to schedule the power system control to achieve operation at a desired security level, while optimising an objective function such as cost of operation. Any specific on-line OPF scheduling function is designed to:

- Operate in real-time or study mode
- Schedule active or reactive power controls, or both
- Achieve a defined security level
- Minimise a defined operational objective.

Most OPF computer programs are able to perform more than one specific function. Most security-constraint scheduling functions are being designed to communicate interactively with the power system operator. In the real-time mode the calculated schedule may be implemented manually or where possible automatically; generations and interchanges can be fed into the automatic generation control system.

3.2.1 Minimum cost of operation

The most critical factor in cost minimisation is the modelling of the generator cost curves. This has not only an effect on overall optimality but is intimately bound with the solution method used. Costs for thermal units are derived from the heat-rate curves, which are often far from smooth or convex.

3.2.2 Minimum active power transmission losses

The controls that can address this objective are all ones without direct costs, i.e. all except MW generations and interchanges. At the same time it should be recognised that some controls, e.g. phase shift transformers and DC-line flows, will not usually be scheduled for loss minimisation because they are more useful for active power control. Thus, loss minimisation is normally associated with voltage/VAR scheduling. It tends to reduce circulating VAR flows, thereby promoting flatter voltage profiles.

3.2.3 Minimum deviation from a specific operating point

The objective function is defined as the sum of the weighted squares of the deviations of the control variables from their given operating values. Such an objective may be used for corrective rescheduling. The main problem is the correct weighting of the individual contributions to the objective function.

It may be suitable to choose the weighting factors on the types and locations of the limit violations such as that of a bus voltage compared to the violation of a branch flow.

3.2.4 Minimum number of controls rescheduled

Some practical approximation to this objective is essential in the many cases where it is impossible to reschedule a large number of controls at the same time. This applies when there are limited means for the automatic dispatch of many controls simultaneously from the control centre, and frequent moving of tapped apparatus is to be discouraged.

3.3 System control

Power system control is an important component of an on-line operation. The overall objective is to control the electric power output of generators so as to supply the continuously changing customer power demand at a given system frequency.

Four basic objectives of power system operation during normal operating conditions are associated with automatic generation control (AGC):

- Matching total system generation to total system load
- Regulating system frequency error to zero
- Distributing system generation among control areas so that net area tie flows match net area tie flows schedules
- Distributing area generation among area generation sources to minimise costs.

The following problems are currently of prime interest in system control:

- Decreasing unit response capabilities due to increasing restrictions on nuclear plant operation
- Higher fuel costs shift the balance of the perceived trade-offs between economy and "tight" control
- Large industrial loads that can change quickly and hence are difficult to follow
- Use of storage devices to improve the system dynamics.

3.4 Restorative control

Power system restoration after major disruption is carefully specified in operator's manuals. The infrequent occurrence makes it difficult to assess the efficiency of pre-specified restoration procedures. In this context the power system training simulator may offer an effective tool for examining current practice.

4 Power system control centre: hardware structure

4.1 Overview

Just as every power utility is unique in its electrical system structure and operating policy, so too are the control centre hardware structures and design considerations. A review of energy control centre systems throughout the world reveals the wide range of hardware configurations currently operating or projected to be in service within the near future. Despite this diversity, two conceptually different and main structures have been developed. Centralised systems consist of a main computer or computers which handle all the SCADA/EMS functions. Distributed systems consist of multiple processing levels; each level is assigned a specific set of functions. This concept enables the assignment of functions to machines best suited to the individual requirements of the respective functions. The main components of the hardware structures are the remote terminals, the communication links, and the real-time computer system. The hardware resources used depend on the following factors: number of measurements and status points, complexity of the automatic generation control, extension of breaker and switching control, scan rates, number of CRT and displays, type of MMI, existence of the mapboard, number and complexity of the advanced power application functions, the existence of a dispatcher training simulator.

4.2 Remote terminal unit

The RTU represents the operator's eyes, ears and hands with respect to the power system (see Fig. 4.2-1). Analog inputs normally come from transducers which generate a volt or milliampere representation of electrical quantities such as kilovolts, amperes, megawatts and megavars. Digital inputs may represent the status of breakers, switches and relays. Digital inputs may also represent kilowatt hours metering pulses to be accumulated (energy values). Analog outputs are intended to pass operating instructions to a setpoint controller (e.g. MW setpoint value for a generator). Digital outputs may command the operation of breakers or may pass operating instructions to a peaking generator unit.

The RTUs always operate from a substation battery (primary power). A minimal MMI is necessary for maintenance purposes.

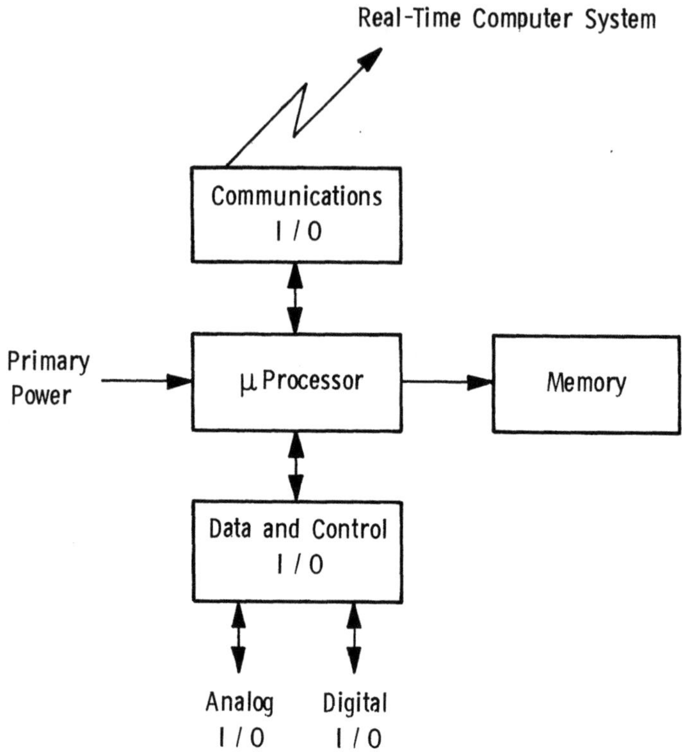

Fig. 4.2-1: The structure of the remote terminal unit

The modern RTU is microprocessor-based, equipped with both programmable read-only memory (PROM), and random-access memory (RAM). Using microprocessors makes it possible to improve the maintenance, to enhance the flexibility and allows performance of further advanced functions. Built-in diagnostic ability is one of the most important features which a microprocessor can provide. The additional features of an intelligent RTU are:

- Report by exception
- Data concentration
- Sequence of events reporting
- Faults reporting
- Digital relaying
- Control sequence.

4.3 Communication

A SCADA/EMS may be separated by many hundreds of kilometres from its power stations/substations with which it must communicate through the RTUs. Therefore a power system control centre has extensive communication requirements. Communication may be performed in parallel or in serial form. Parallel communication can be performed at a very high speed, is relatively expensive and is mainly used within computers. Serial communication is performed using asynchronous or synchronous transmission principles. Asynchronous transmission means that the transmitter and the receiver are not synchronised all the time. When synchronous transmission is used, the transmitter and the receiver have clocks that are synchronised all the time. Data can be exchanged between equipment using the following transmission principles:

- Simplex
- Half-duplex
- Full-duplex.

The communication media options are: metallic cable pairs, powerline carrier, microwave fibre optic cable, satellite communication. The evolution of communication systems from traditional analog to digital mode continues. This trend is the result of the known advantages of coded digital signal transmission: transmission errors can be reduced or even corrected by means of suitable correcting codes. An integrated digital network can fulfil the following requirements: telecontrol, teleprotection, telephony, data communication, text communication, facsimile, video, radio telephone.

4.4 The real-time computer system

4.4.1 Central processing unit

The real-time computer system shown in Fig. 4.4-1 is the most important part of the network control system and comprises central process unit, main memory, mass memory, equipment for man-machine interface, and peripheral equipment. Some of the features of the real-time computer systems are: low memory cycle times, multiple external and internal interrupt levels, fast access disk, multiport memory banks, direct memory access, internal real-time clocks.

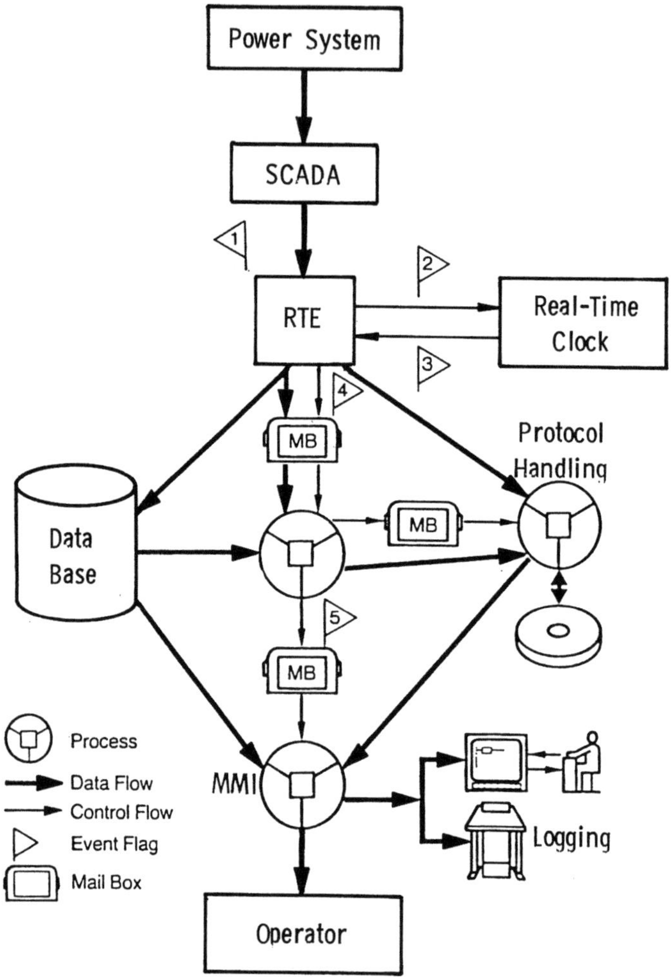

Fig. 4.4-1: Real-time computer system for power system control

Table 4.4-1 shows the frequency of different energy control functions together with the approximate CPU utilisation in percentages. It is clear that the average CPU time requirement depends very much on the CPU time per execution for each monitoring and control function. This is not only a hardware aspect because software engineering also plays a major role in the relative performance of CPUs.

Table 4.4-1: Approximate CPU utilisation of energy control functions

Function	Frequency	% CPU Utilisation
SCADA	2 s	6
AGC	5 s	
ED	10 min	
Man-Machine Interface	5 s	15
Disruption Data Logging and Storage	10 s	1
Real-time Network Analysis	15 min	10
Security Analysis	15 min	15
Optimal Power Flow	24 runs/day	3
Unit Commitment	3 runs/day	1

Extensive studies show that for a SCADA/EMS there is no single best choice; no one CPU available today can satisfy the full set of SCADA/EMS requirements. Only a combination of mainframe, minicomputer and microcomputer meets all objectives. At first glance, this practical conclusion contradicts Grosch's law known since 1953. This law states that the cost of a computer system increases as a function of the square root of the computer power and that the cost of a given computer job will decrease by utilising more powerful computers. This means that the most economical solution is to procure the largest computer compatible with the user's particular needs and that one large computer is to be preferred to several small ones. It follows that as long as Grosch's law holds, no convincing argument can be made in defence of a mix or a distributed computing architecture.

Re-examination of Grosch's law for the SCADA/EMS industry leads to the conclusion that computers can no longer be regarded as one homogeneous product and that they should be divided into the following categories for analytical purposes: microcomputers, minicomputers, small mainframes and super computers. Within this classification Grosch's law seems to hold good for each of the four above defined computer categories. However, a differentiated computer approach and the utilisation of different computers for various SCADA/EMS functions is a necessity for both the suppliers and for the electric power utilities.

4.4.2 Computer memory system

The memory system of a power system control computer is organised in several levels depicted in Fig. 4.4-2. Starting from the bottom, at the lowest level, there will be an archival storage, frequently called tertiary storage, usually comprising a set of magnetic-tape units. The next level upwards, the secondary storage, consists of magnetic disks or drum memories or slow magnetic-core memory units; the secondary storage is called the backing storage. The memories of the uppermost level, comprising the primary storage, are housed in and functionally belong to the central process unit. The primary memory is operationally distinguished from the other storage devices by the fact that it is the part in which the memory-reference instructions find their operands on the basis of addresses.

A very important concept characterising the real-time computer system is that of virtual memory and cache memory. The principle of virtual memory is commonplace in modern computers and has been found advantageous mainly from the programming point of view. When designing a computer architecture, one of the basic features is the address capacity of machine instructions, i.e. the range of absolute addresses in the memory system which can be directly indicated in a machine-code instruction (direct addressing). The idea applied in virtual-memory computers is to make this range fairly large, corresponding to that of the backing memory. Another type of virtual memory which seems to have become accepted even in small computers, primarily due to developments in semi-conductor component technology, is the cache. This is a buffer which works according to the same principle as the virtual memory but which is mainly intended to speed up the access time of primary storage. By virtue of the cache and virtual memory concept the computer may seem to have a one-level memory with the capacity of the secondary storage and speed which is essentially the same as that of primary storage.

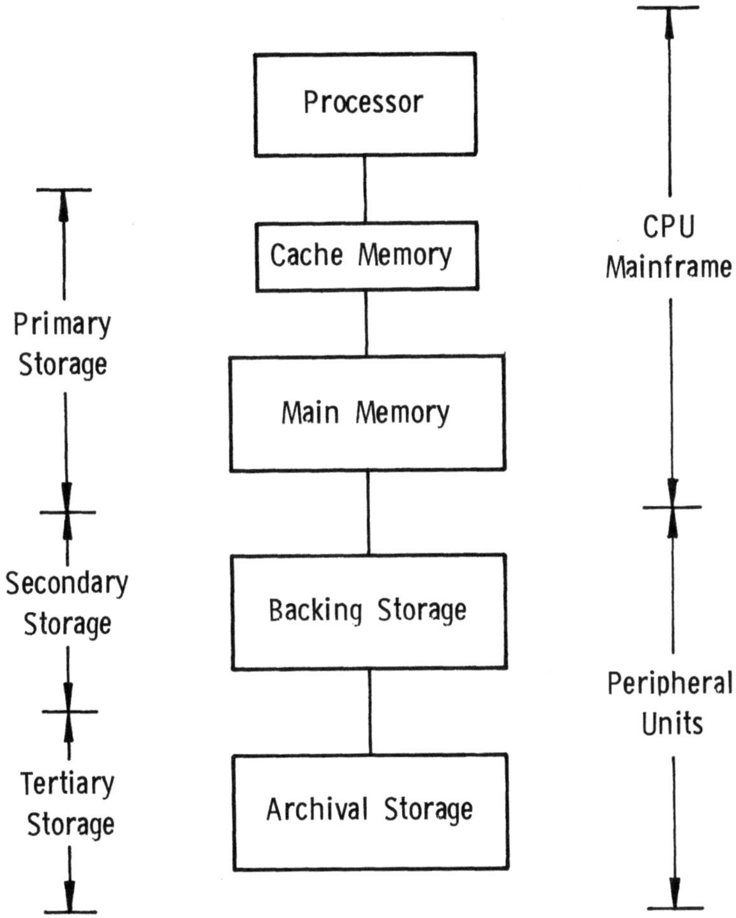

Fig. 4.4-2: Memory system of the real-time computer system

4.4.3 Man-machine interface (MMI)

The man-machine interface comprises console equipment with cathode-ray tubes (CRT), mimic boards, chart recorders and other peripherals. The dynamic wall display is intended to give an overview of the power system. This is best accomplished by a simplified representation preserving the geometrical orientation of the system as much as possible. The wall diagram must not be overcrowded with details; details of the network diagram are best left to the CRTs. There are different opinions on the need for dynamic wall display at control centres. It may be considered as an option.

Mixed arrangements are also used: some information (e.g. switch status) fed directly from the RTUs, or some information (e.g. measurements) fed from the computer. In addition to showing the overall network topology, the dynamic wall diagram can also display the loading of the network or some important alarms. The wall diagram can be driven by the main computer or by the front-end computers. In the latter case, the wall diagram can serve as a back-up for the main computers.

For the operation of the dynamic wall diagram there are two basic approaches:

- The dark-board operating approach
- The light-board operating approach.

Using the dark-board operating method, except for the generator status, only abnormal status of equipment is indicated by lighting that appropriate symbol.

The CRT display has become a universal feature of power dispatch control centres. The CRT display can provide practically all the necessary interactions with the human operator.

The full-graphic CRT is becoming the state-of-the-art in network control centres. The experience of the movie industry shows that a full graphic system requires a very powerful machine and a large memory. For example, for only one frame of animation, 10 minutes of CRAY CPU computer time would be necessary (for 1 second of film approximately 20 frames of animation are necessary). In order to obtain a high quality resolution of about 2000 x 1500 at 24 bits per pixel, an amount of up to 8.5 Mbytes per frame are necessary.

An effective man-machine interface must have a response time of about 2 seconds on the average, from the instant the operator makes a CRT selection up to the time requested display is completely on view. The parameters that characterise the responsiveness of the MMI system are defined and specified in a quantitative manner in Table 4.4-2.

Table 4.4-2: Man-machine interface characteristics

Feature	Definition	Time Requirement
Response Time	Time from dispatcher request until the required display request is acknowledged	0.5-2.0 s
Update Time	The time interval at which the dynamic data on the display is renewed from the data base	3.0-5.0 s
Refresh Time	The time necessary to write the CRT screen completely from CRT memory	10.0-20.0 ms

4.5 Review of hardware structure for network control centres

A variety of different computer configurations is currently used in network control centres. These configurations have evolved over a period of time in response to changing hardware availability, costs and changing control centre functions. All have a common attribute: they are designed to provide both restart and back-up capability in case of a system failure. Because of the variety of configurations, the following classification scheme is helpful:

- Dual computer configuration
- Front-end computer configuration
- Triple configuration
- Quad computer configuration
- Distributed computer configuration.

Two general approaches are considered: loosely and tightly coupled systems. Loosely coupled systems transmit data from one system to another through the use of a communication link or dual-access input/output devices (a dual access disk). Tightly coupled systems use high-speed shared memory to transfer data.

4.5.1 The dual computer configuration

The most common configuration found in energy control centres is the dual computer configuration consisting of identical computer systems operating in "primary" and "stand-by" modes. Each side of the dual configuration may comprise one or more computer systems performing the assigned functions. Normally, the primary side of the configuration performs the real-time functions of data acquisition, control, man-machine interface, security assessment, etc., while the stand-by configuration is ready to assume the primary role when required. To support the stand-by role adequately, a linkage, referred to as coupling, is available. The use of a single computer for the primary control system with a duplicate for the back-up is termed a dual computer configuration. As shown in Fig. 4.5-1, the entire configuration is duplicated including the input/output hardware. A switching mechanism permits the use of either machine as the primary. A loosely coupled dual configuration communicates through a data communication link at relatively high speeds or by accessing-dual-ported disks.

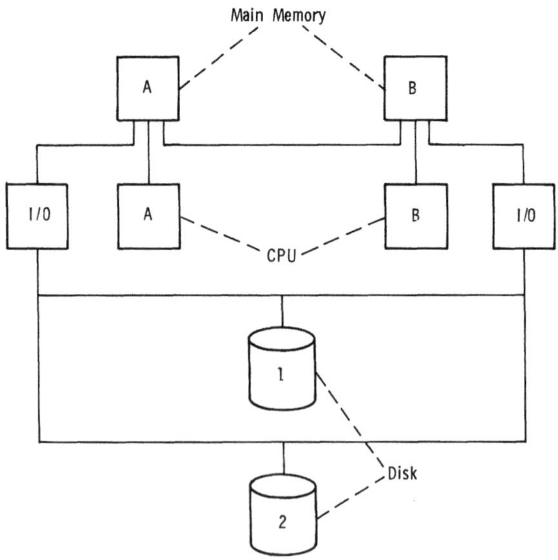

Fig. 4.5-1: Dual computer configuration
A: primary computer system
B: stand-by computer system

Tightly coupled dual configurations (Fig. 4.5-2) share a common memory bank. The shared memory is like a dual access device in that it is capable of being switched in or out of either processor.

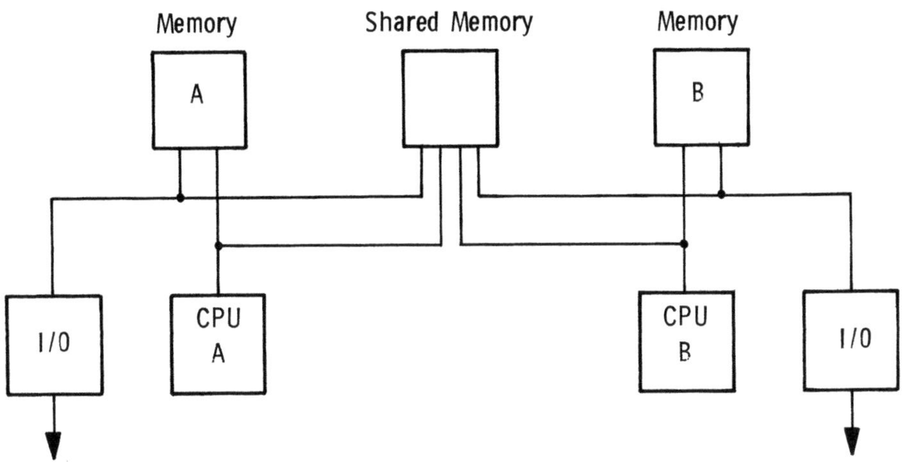

Fig. 4.5-2: Tightly coupled dual computer configuration

4.5.2 The front-end computer configuration

The dual configuration, though frequently comprising only two processors, can be set up with four or more processors. One such configuration is formed by the addition of front-end processors to the dual computer system (Fig.4.5-3).

Fig. 4.5-3: Dual front-end computer configuration

The front-end processors normally perform the data acquisition and control functions for transference to the main computer through a loosely coupled high-speed data link. Front-end computer configurations link special processors to the control computers to off-load the data acquisition and associated processing functions from the higher level control computers. Each front-end processor is coupled only to its corresponding higher level computer. The functions assigned to the front-end computer may vary from simple communication interfacing through to sharing of the control tasks. If the front-end computer performs all scanning and communication with the remote terminal units but with no data reduction, it would simply transmit the raw data collected to the higher level computer and receive process outputs from the higher level computer for transmission to the RTUs at the same rate. Additional functions such as limit checking and alarming may be assigned to the front-end computer. Coupling between the front-end computer and the higher level computer can itself be loose or tight.

4.5.3 Triple configuration

Another method of addressing the problems associated with the reliability of the dual configuration is through the use of a "triple" configuration. This configuration is composed of three identical systems coupled by redundant high-speed data bases or shared memory. The third system could be configured independently of the other two and manually connected. During normal operation of all three processors, the third system would be available for off-line studies and software development, and would provide back-up support during normal system maintenance.

4.5.4 The quad computer configuration

One pair of front-end computers flexibly coupled to another pair of computers for higher level control functions is called a quad configuration (Fig. 4.5-4). The quad computer configuration differs from the front-end configuration in the fail-over combinations available. Front-end configuration systems pair a specific front-end computer with a specific higher level computer. When failure occurs, all functions are switched to the other pair. Hence such a system is totally disabled if a failure occurs in a front-end computer of one pair and the higher level computer of the other pair. In contrast, a quad configuration permits separate switching of the front-end and higher level computers so that either of the front-end computers may be used with either of the higher level computers. Coupling among processors may be tight or loose.

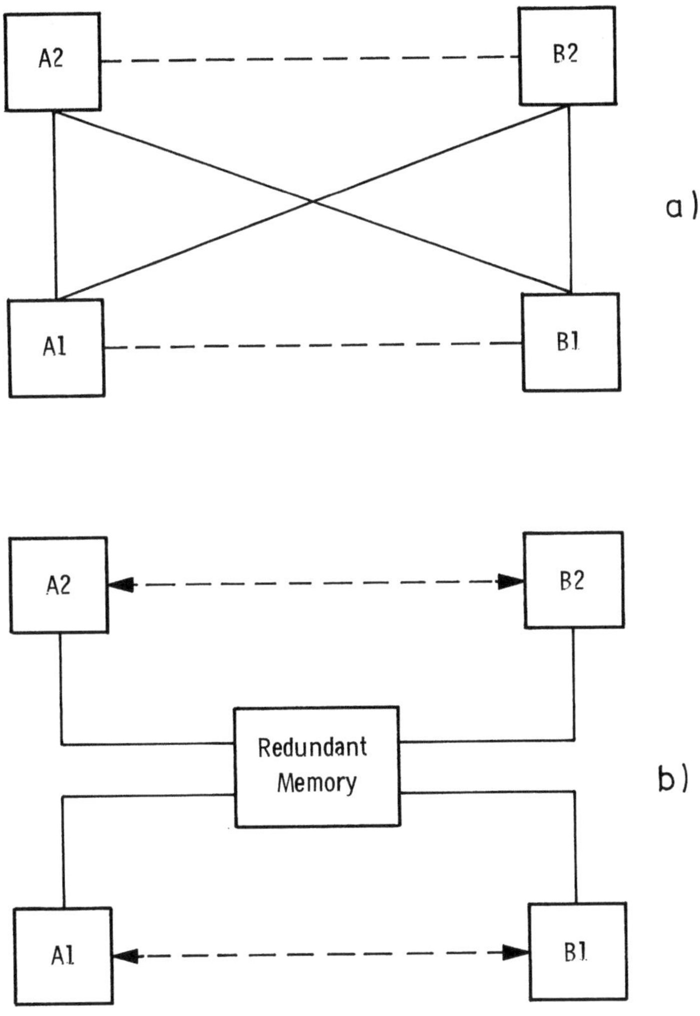

Fig. 4.5-4: Quad computer configuration
a) loosely coupled
b) tightly coupled

4.5.5 Distributed system configurations

Both the front-end and quad configurations assign different functions to different processors and hence distribute the control centre functions. A distributed system, as shown in Fig. 4.5-5, is one with multiple processors coupled through a communication system and sharing I/O processors and devices.

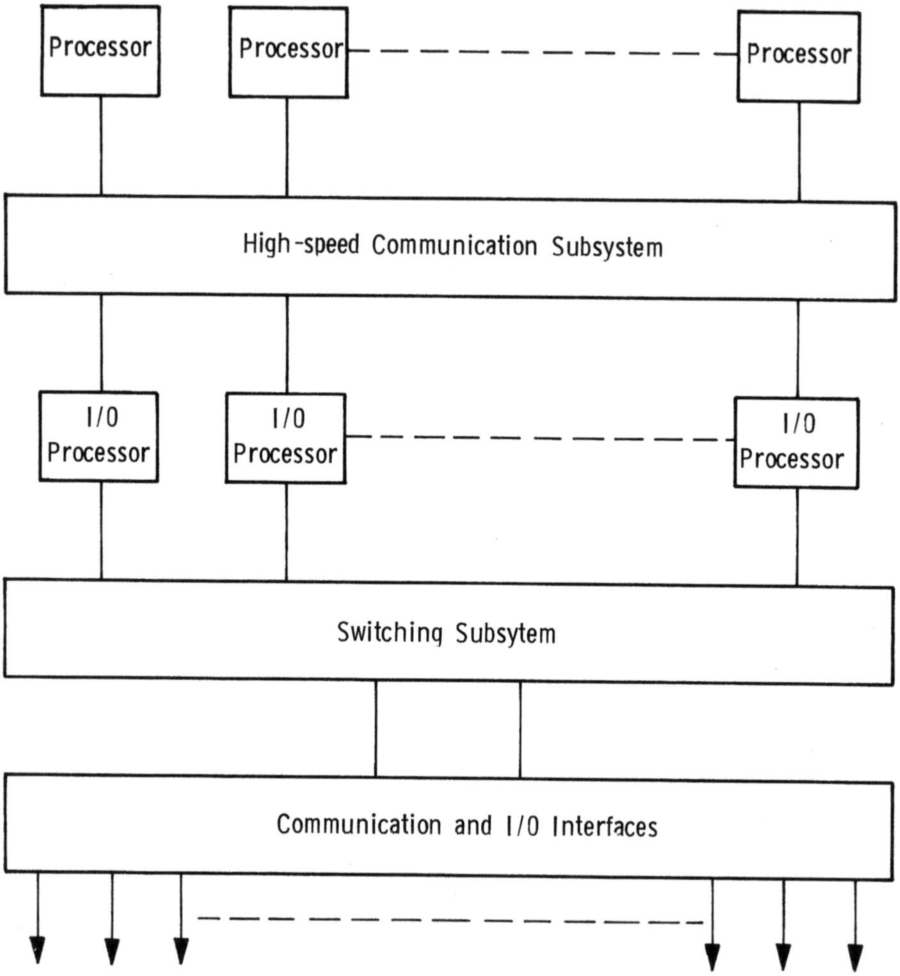

Fig. 4.5-5: Distributed system configuration

In addition, the operating system software is designed for the system as a whole rather than for individual computers. The underlying concept is to assign different functions to different processors and to operate them concurrently. Communication among processors is through high-speed data links, special data buses, or shared memory. Distributed systems can be divided into the following categories:

- Dual distributed configurations
- Degraded mode-back-up with redundant communication links
- Non-stop configurations
- Cannot fail configurations
- Redundant hardware configurations.

4.6 Hardware design considerations

A network control centre consists of two basic subsets of functions:

- SCADA functions
- EMS functions (advanced power applications and dispatcher training functions).

These subsets impose different requirements on the hardware. SCADA needs frequent and fast context switching from one program to another and a great number of logical operations. The SCADA functions are characterised through tight real-time requirements; they are interrupt-driven, prioritised and repetitive.

EMS imposes a high burden concerning numerical, floating-point operations on large amounts of data. Different functions have various requirements concerning the hardware; these justify a differentiated multi-computer approach. No one computer architecture available today can satisfy the full set of SCADA/EMS computing loads.

The hardware designer must therefore look at structures combining mainframe, minicomputer and microcomputer architecture that could respond to both large numbers of external interrupts associated with the necessary SCADA requirements and the heavy arithmetic computations required by advanced power applications and training simulator functions. Presently a large network control centre requires a total computing load of about 13 MIPS, an arithmetic computing load of approximately 4 MIPS, 1000 external interrupts per second, an I/O load of about 2000 kilobytes/second and 17 000 data-based transactions/second. In order to allow further expansion the following requirements must be met: 37 MIPS for the total computing load, 13 MIPS for the arithmetic computing load, 3000 external interruptions/second, approximately 6000 kilobytes/second for I/O load, and 50 000 data-based transactions/second.

An interesting approach to the hardware structure is the so called "mix-and-match" structure. Rather than relying on a symmetrical structure of mainframes or minicomputers, a mixed structure of micro-, minicomputers and mainframes is used. The idea behind this is to match each specific control function to some specific hardware features. By using more than one type of computer the overall output is faster. The SCADA system places the most stringent requirements on computing resources because it includes many small programs and large streams of real-time raw data. Analysis shows that the required instruction mix is heavily weighted towards logical operations and could be accommodated efficiently with a 16 bit instruction set.

The real-time aspects of data acquisition programs demand the rapid processing of many external interruptions. The advanced power applications are less demanding as real-time concerns but they require extensive arithmetic processing and a computer that can accommodate large programs. The power applications are best housed in computers with a large set of floating-point instructions, the precision requirements being confined within 64 bits. Computing speed is also of primordial importance. In one supplier's concept, the real-time computer architecture is built around four processing levels:

- Advanced Power Applications - processor
- SCADA - processor
- Data Acquisition - processor
- Man-Machine Interface - processor.

The data acquisition processor provides the high interrupt activity level and high transaction rate required for a communication environment. The MMI processor provides the lower interrupt activity level, as well as the high transaction rate required for displays and console processing. An IEEE Task Force has pointed out that a major problem concerning the power system control centre hardware design is the lack of ability to upgrade a system easily. In planning for the expected life of a control centre, it is important that the hardware can accommodate horizontal growth (more RTUs, CRTs, more data, more displays) as well as to a vertical growth (new and more complex SCADA and EMS functions)

4.7 Hardware obsolescence

The obsolescence of SCADA/EMS is a growing problem which most utilities will face during the coming years, some for the second time. In the computer industry, life cycles are relatively short and they tend to get even shorter. This trend is related to the high rate of innovation and increased competition. The short 5 - 10 year life span of computer systems is a fact of life that utilities find hard to accept; they are used to 40 year life spans for power plants and network elements. By the time a SCADA/EMS is 10 years old the associated computer family will have evolved through several generations of hardware. Spare parts and maintenance for the old equipment can become very expensive and, in some instances, unavailable. An associated area of concern is the fact that some suppliers do not offer continual upgrades of their products. There is a gap between the life expectancy of the SCADA/EMS and the time span in which suppliers of computer hardware are able to offer economical and available equipment support.

4.8 Performance of SCADA/EMS

Today, there is a lack of ability to measure, predict and control SCADA/EMS performance. Unresolved problems in these areas are indicated in the following observations:

- There are no standards or industry-accepted benchmark programs to measure the SCADA/EMS performance
- There is disagreement between utilities concerning adequate SCADA/EMS response times
- It is difficult to develop a realistic worst case scenario (computer resource/cost trade-off)
- There is a need for accurate means and tools to predict and measure SCADA/EMS performance.

Based on existing experience, the following guidelines for control centre hardware design are recommended:

- A CPU utilisation time of 75% during normal operation should be considered as maximum design limit.
- A 90% utilisation of the mainframe memory for resident and 95% for overlay areas should be considered as maximum design limit.

- For systems using virtual or memory mapping techniques a design limit for context switching delays (roll in/roll out) is set at 1% of the total CPU computational time.
- Data acquisition shall use no more than 40% of the system's computational time.
- The peak loading, which is determined by the time required to communicate with all remote terminals connected to it under worst case system conditions, shall not exceed 95% of the total time during any five minute period.
- The maximum design limit for utilisation of the I/O channels is 90% of the total capacity.
- Maximum design peripheral service time is 10% of the available CPU time.

System response delays in the MMI should be limited to 2 seconds for an operator action. Response is defined as the time between operator action and acknowledgement of the action by the system. The maximum utilisation of real-time auxiliary bulk memory must allow a reserve area equivalent to two times the system mainframe memory size. For redundant systems, the minimum design level of availability is 99.95%. The utilities have also addressed the problem of SCADA/EMS loading and the following activity states are defined. Fig. 4.8-1 shows the CPU performance requirements corresponding to the following three activity states.

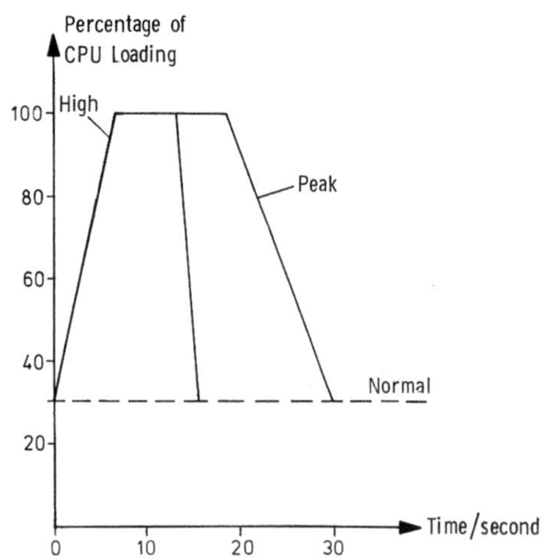

Fig. 4.8-1: The performance requirements for the CPU corresponding to the three activity states

The normal activity state corresponds to the typical condition of the power system. The hardware configuration is normal with all remote terminals and peripherals connected; all the programs are running. The power application functions are scheduled as shown in Table 4.8-1. The average scheduling and run times vary in a considerable time range. The values given in Table 4.8-1 may be considered representative for many EMS installations. However, different values may be observed in specific installations depending on the network size, the hardware technology and the software implementation.

Table 4.8-1: Run times in the normal activity state

Function	Average Scheduling	Average Run Times
Automatic Generation Control	2-3 s	0.5 s
Economic Dispatch	5 min	10 s
Generator Reserve Monitoring	2 min	5 s
Interchange Transaction Evaluation	1 h	30 s
State Estimation	10 min	30 s
Operator Load Flow	On Request	10 s
Optimal Load Flow	30 min	60 s
Contingency Analysis	15 min	20 s
Short Circuit Analysis	0.5-1 h	10 s
Load Forecasting	4-8 h	20 s
Unit Commitment	1-4 h	120 s

Additionally the data acquisition system updates the indications, measurements and energy values. This may be done by using spontaneous reports by exception or cyclic processes. The normal activity state is defined as follows:

- One percent of all status indications change every hour; each indication change is logged in an event/alarm list
- One percent of all measurements change every second
- Each minute, 0.01 percent of these measurements exceed the limits and are logged in an alarm list
- One percent of all calculated values is updated every second
- The operator calls for a new display at a rate of one picture per minute per console
- The operator changes a set-point or changes a value once every 5 minutes per console
- The operator sends an open/close breaker command every 15 minutes.

The high activity state results as a consequence of a major power system disruption such as:

- Loss of the largest single generator
- A short circuit at a major power generation or transmission station; the fault is cleared by primary protection
- A local thunderstorm resulting in tripping/reclosing of a number of network elements.

The level of high activity state is defined as follows:

- Status change (5 - 25%) of the total number of indications
- Protection alarms (2 - 6%) of the total number of alarms
- Limits violations (7 - 25%) of the total number of measurements.

The peak activity state results as a consequence of a very severe major system disruption such as:

- A system blackout
- A three-phase short circuit at the largest power or transmission station
- The primary protection having failed.

The level of peak activity state is roughly defined by a change to zero of all measurements and a change in the status of 30% for all breakers. The performance requirements for the CPU loading, averaged over a 5 minute interval, should, in a normal activity state, not exceed 30% (all critical functions are running). In the high and peak activity states the system must survive without loss of data and return to normal within the periods of time shown in the Fig. 4.8-1. During the high and peak activities the power application functions and the non-critical software processes are inhibited.

5 Power system control centre: software structure

5.1 Overview

It took more than 20 years for terms like SCADA, EMS, etc. to gain acceptance. As more utilities are using modern control centres, new functions are emerging. At the same time, under the same connotation, "old" functions are growing in complexity.

These changes have meant that the following slightly modified software classifications have had to be introduced:

- Data acquisition system
- Supervisory control system
- Real-time software environment
- Data base management system
- Man-machine interface
- Executive decision support system
- High level decision support system
- Inter-utility communications.

All these software subsystems are intimately interwoven. The control centre software with its multiple interfacing and complex intrinsic relationship is represented as a layer model in a simplified manner in Fig. 5.1-1.

The bottom level of software is the system support software, including the operating system, I/O drivers, telecommunication software, inter-computer communication software and diagnostic software. These items interface directly with the hardware.

The second layer uses the system software to support the basic applications. It performs the generalised functions such as data management, supervisory control, data acquisition, man-machine communication and basic applications functions.

The executive decision support system is the third layer of software. It is a power application software that uses the data gathered by the second layer and accomplishes the same monitoring and control functions desired by the dispatcher.

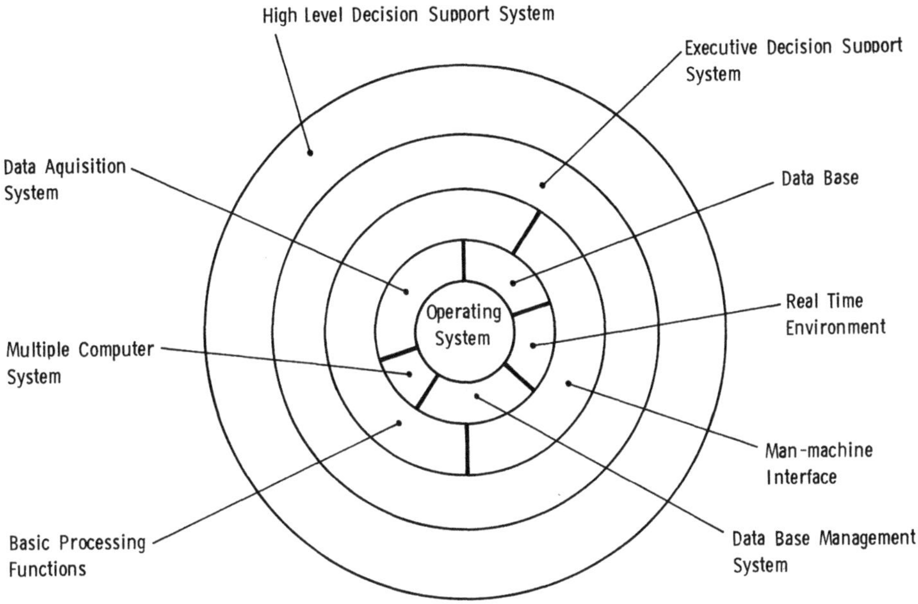

Fig. 5.1-1: Software structure of the network control centre as a layer model

The high level decision support system is at the top of the software structure. This software provides the on-line network analysis, study network analysis and other complex control and economic dispatch analysis.

5.2 Data acquisition subsystem

The DAS transfers the data from the remote terminal units (RTU) to the control centre and vice-versa. The DAS supplies a number of functions such as:

- Collection of data
- Error checking and plausibility tests
- Conversion to engineering units
- Limit checking
- Handling of alarms.

5.2.1 Collection of data

There are two standard techniques for data acquisition. The data scan techniques retrieve all data from the remote terminals at a rate determined by their relative importance, regardless of whether the data has changed. The normal scan consists of interleaved two-second and ten-second scans. All status and key analogues are retrieved as a part of the ten-second scan while the remaining analogs are retrieved throughout the ten-second scan period.

In the transmission-by-exception technique the data are received by tranceivers and compared with the stored data value for each of the points. If there is no change in status points or if all analogue values are within the specified deadbands of the stored values, no further transmission is requested. Normally the electric power system is in a quiescent state. Therefore transmission by exception results in a better overall system for the same amount of hardware resources. While transmission by exception sharply reduces the communication loading under most conditions, it can cause heavier loading than the periodic scan technique under emergency conditions.

5.2.2 Error checking and plausibility tests

If the tranceiver has a communication failure, it automatically tries to process the failed message again a specified number of times. If after all attempts communication is not established, the tranceiver returns a communication failure indication which includes the reason for failure. As soon as communication is re-established, all points at the RTU are automatically retrieved. This procedure assures that the data base is up-to-date as soon as communication is re-established with an RTU.

5.2.3 Conversion to engineering units

The retrieved values are converted into engineering units and replaced in the data base. First order and fifth order polynomial conversion from A/D-values to engineering units are provided.

5.2.4 Limit checking

The engineering unit values are checked against the limits established as reasonable for the point. If these are exceeded, the data quality is marked as "wrong data" and there is no further processing of the point. The point could be marked for inclusion in the post disturbance review. The limit checking function translates an analogue value into a three state status value: out of limits-high, in limits and out of limits-low. The analogue value to be checked is compared against the limits and the appropriate status value is generated. Any number of limits can be applied to an analogue value.

5.2.5 Handling of alarms

When an alarm is detected, the following sequence of events occurs:

- An alarm message is added to the log
- An alarm line is added to the alarm summary by category
- The alarm appears on the related display
- An audible alarm sounds
- The point is marked as having an unacknowledged alarm.

5.3 Supervisory control subsystem

The supervisory control commands generally begin in the man-machine subsystem. Exceptions are commands issued by automatic generation control or load shedding functions. Once a complete command sequence has been fulfilled, a composite command is sent to a control module that checks control feasibility and then translates the command into a control sequence for use by the RTU communication subsystem.

If all permissive conditions are met, the control command is sent to the RTU communication subsystem. This subsystem interrupts the current scan to send out the control. One of three responses are returned:

- The select was unsuccessful (after "n" retries)
- The execute was unsuccessful
- The command was transmitted successfully.

If the command was unsuccessful, the operator would be informed of the command failure. If the command was unsuccessful for any reason, the attempted command is logged with the reason for failure.

If the command was successfully transmitted, a timer is started to verify actual execution of the command. If verification is received before the timer has attained the allocated time, the operation is logged as successful.

Otherwise the operator is informed that the operation cannot be automatically verified.

5.4 Real-time software environment

5.4.1 Operating system

A real-time operating system includes some distinguishing features such as:

- Additional time management facilities for scheduling synchronous tasks
- Extensive multitasking capabilities
- Extensive interrupt capabilities.

5.4.2 Real-time traffic manager subsystem

This subsystem schedules and performs the supervision of computer processes, manages the computer resources, allocates the different peripheral devices, checks for the existence of conflicting tasks and manages them. Other central functions are error-handling, error logging, and the management of the multiple computer configuration. In the case of main computer failure, this subsystem transfers the results to the hot stand-by computer or automatically restarts the critical real-time functions. After a failure, errors can be traced back with the program and the system state software context areas of main memory will be dumped on disk for later diagnosis.

5.5 Data base management system

5.5.1 Overview

A data base is, from a logical point of view, a model of the knowledge structure of the power utility, as it exists to support the operating and planning activities. The data base is a common body of data that is used by many different SCADA/EMS functions. The data base is the core of a computer-based control centre. All the data required by the control centre software and by the operator are located in the data base. In a power system environment, a significant proportion of the data requires frequent updating. It is often necessary to expand an existing data base. Snapshots of the data at a certain time are needed to ensure that the data used by the application programs are consistent. A well designed DBMS will improve the overall system characteristics. However, a DBMS imposes a heavier load on the hardware and causes a more complex and expensive software.

In a modern SCADA/EMS all interfaces among programs and all interfaces between users and programs pass through the data base. Programs do not contain I/O to user devices; displays of results are extracted from the data base.

5.5.2 Requirements

Some important attributes of a modern DBMS are listed below:

Integrity: data must be accurate and consistent. This means that range, validity and plausibility checks on the data should be applied to data entering the data base.

Resiliency: data must not be lost or destroyed. Errors must be detected and corrected.

Reliability: the relationships between and among data must be well understood, documented and defined. A DBMS must support the creation of new relationships.

Shareability: support software must provide simultaneous sharing of data resources, while optimising their use.

Security: the architecture must provide appropriate security controls to defend the system against malicious mischief.

Independence: the DBMS should permit the separation of programs from the logical and physical data structure. As a consequence the programs can be changed independently from data structures and vice versa.

Accessibility: this attribute means the facility to handle the data base by non computer orientated people.

Performance: a data base should be tuned as the requirements change.

Administration: the DBMS includes tools that allow the data base resources to be managed.

5.5.3 Software aspects

The DBMS is basically a software package within the control centre software system. It supervises the processing of the data base functioning. Some of the DBMS functions are:

- Controlling the physical layout of the data in the data base
- Updating the data elements
- Interfacing with the application programs.

The DBMS covers all off-line and on-line data handling activities:

- Off-line data preparation allows for plausibility tests of static data used by all software functions, e.g. rules of the topology description, range of the limits, range of impedances values
- On-line data allocation and storage is one of the most critical functions.

With the processing in control centres becoming more modular and decentralised in nature, much of the DBMS could be transferred to a dedicated processor that functions similarly to a front-end processor for communication. The host computer will interface with the data base processor. A dedicated data base processor permits the use of specialised hardware and software because the processor will be restricted to data base functions. Therefore the data base processor does not need all the characteristics of a general purpose machine, since such a processor would not require either floating point instructions or fast arithmetic operation.

There are also some disadvantages; one of which is that it may be more costly. Another could be a time response penalty if the interface between the host computer and the data base processor is not properly designed.

5.5.4 Structure of the data base

A survey of the data base technology reveals the following three logical structures:

- Hierarchy (tree)
- Network
- Relational.

The layout of the data base has a great impact on the runtime performances of the different application programs. In a relational data base, data is logically arranged as a collection of two dimensional arrays. Each of these arrays is called a "data type" or a "data entity".

A data entity is an object, an event or an abstract concept about which information is retained. Examples would be a transformer, an alarm or a feeder. Each data type or data entity has pieces of information associated with it. For a transformer these might include grid location, voltage rating, reactance, etc. These pieces of information are called data elements or attributes. The relational data base structure is based on arrays and attributes. The arrays coincide with the types of network components and the attributes represent their various characteristics. (Fig. 5.5-1).

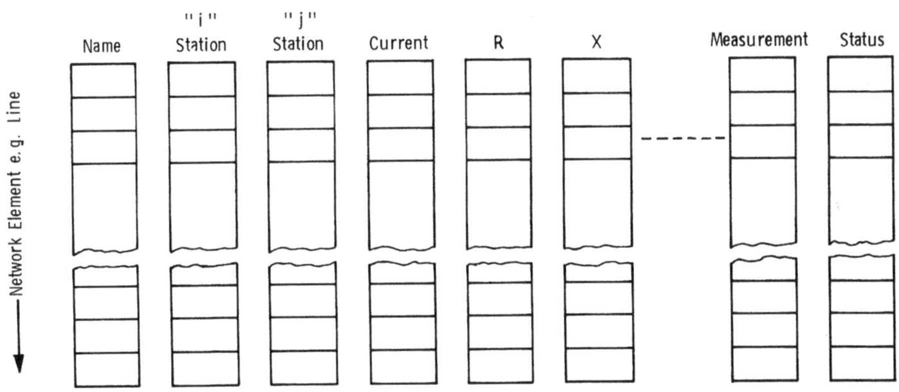

Fig. 5.5-1: Data base structure

The "view" of the data base that a specific function required is called a data base subscheme diagram. This diagram represents the structure of the data for the specific function (e.g. on-line load flow). At the same time a data base subscheme diagram contains the data requirements of the function.

The data base subscheme diagrams are part of the data base master diagram. This master diagram or scheme will combine all the function subschemes into an overall SCADA/EMS view of the data. This master diagram constitutes the logical structure of the data base and represents the logical data base design. The logical data base design is the definition of the structure of data, independent of individual applications and of hardware and software mechanisms used for its implementation. The next step is the physical data base design which defines the method for storing the content of the data base in a specific hardware and software environment. Physically, the data base consists of data arranged in high- and low-speed storage media. Residing in the high speed main memory are the data coming in real-time with a high frequency of access. In a power system environment, a significant amount of the data requires frequent updating. Data residing in slower auxiliary (mass) memory is generally network data, constants, variables and historical data that is infrequently accessed.

5.5.5 Storage and effort requirements

The data base is a vast storehouse of static and dynamic data items. Figure 5.5-2 shows the amount of static data items versus the number of nodes for a control centre with advanced power application security functions like topology, state estimation, operator's load flow, contingency analysis. It can be seen that for a 500 bus network, the data base contains more the 0.5 million data items.

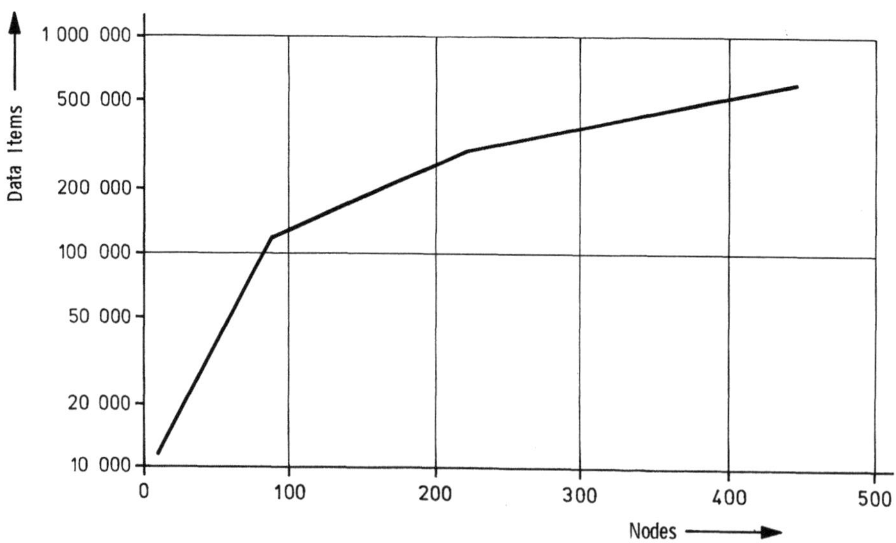

Fig. 5.5-2: Amount of static data items versus the number of network nodes for a control centre with advanced power application functions

For a typical control centre with SCADA and EMS functions and a 380/220 kV electrical network containing 150 buses the data base would contain about 1.5 million static and dynamic data items.

Corresponding with the growth of data items there is a growth in the necessary memory. For a 250 bus network, the amount of memory for the static data necessary for the security functions attains 1 MB. Due to the necessity of handling many data entities with complex relationships between these data entities, the compilation of a data base for a SCADA/EMS system corresponding to a network with approximately 500 nodes is a very demanding and work-intensive task.

5.5.6 Access methods

For real-time application programs the access method is of primary concern. The data base user can access one or many data items with a single call. This call can request any logical combination of attributes. Normally, a data base access relates to the data in a single data type. However, special features allow access to data from more than one data type in a single call. The data base call should indicate the attributes to be accessed. The access can be by attribute names. For faster performance, the data base call can be fully pre-defined at system-build time and the access information saved for use in real-time. Typically, projects require about 3000 real-time values and 3000 status indications updated every 2 seconds. The data base is not only a vast, highly organised storehouse of information, it is also a superior means of communication between functions. For a typical project, the data base serves more than 400 separate programs and load modules for interprogram communication. About 30 colour displays are also updated every 2 seconds.

5.5.7 Performance requirements

A typical DBMS should fulfil the following requirements:

- One second response to operator request
- Two seconds scan cycle
- Two seconds display update.

The availability of the DBMS must be better than 99.9%.

5.6 Man-machine interface

5.6.1 Importance

The success of a SCADA/EMS based control centre is due in part to how well the MMI is designed. Are the displays understandable? Are the displays consistent from function to function? Can data be easily accessed, entered, and is it presented in a clear and concise manner?

The MMI strongly influences the "operability" of a network control centre, i.e. the ability of the power system to be operated safely and efficiently. In the early 1950s the SCADA systems were a great leap forward compared with electromechanical/analogous/telephone techniques but still simple enough for the operator to cope with the problems of monitoring and control. Now in the 1990s the available systems have moved towards very sophisticated SCADA/EMS based control centres. The problem is that such systems have often exceeded the abilities of many operators.

The modern control centres provide vast amounts of data and information and may compound rather than solve the operator's problems. The problem becomes even more critical in an emergency situation.

5.6.2 Human factors

Electric utilities constitute one of the most demanding industries in relation to human factors. The high requirements concerning CRT displays are based on the fact that the dispatcher remains, as previously, the key element in the power system operation and operational planning.

Power plant and power system experience is full of incidents in which - because of the vast amount of data and poorly designed and overcrowded displays - the operational staff cannot determine the overall situation fast enough, i.e. the operators concentrate too much on examining "individual" data and are losing perspective of the general process. It means that good information, poorly presented, promotes lack of understanding. The advent of CRT-based displays has not been the panacea that was promised. While these systems have brought significant amounts of new data to the system operators, they have in some cases made the operator's job more difficult due to confusing presentation in which logical relationships between values may not be apparent.

The one-line diagrams displayed on CRT show the station layout or the network topology with power system data and allow control/switching operation of system equipment. These one-line diagrams provide too much data on a single display. The attempt to represent too many details hampers the capability of the display to represent the station or network structure adequately. The man-machine interface becomes even more critical in an emergency.

The MMI designers need to be especially motivated by the recent questions which arose in the power industry through such events as the Three Mile Island (TMI) incident. At TMI what began as a relatively mild operational transient resulted in almost total destruction of the reactor core due to the operator's inability to determine plant conditions from displayed information.

Another important ergonomic consideration is the rate or the speed of feeding information back to the operator. The display should be in accordance with the operator's speed and capacity to interpret the complexity of the displayed information.

5.6.3 <u>Display formats</u>

This section presents the available display formats. The following is a list of such displays:

- Overview one-line diagrams
- Substation one-line diagrams
- Network tabular displays
- Substation tabular displays
- Transmission overview one-line diagrams
- Subtransmission overview one-line diagrams
- Distribution overview one-line diagrams
- Alarm summary display
- Clearance and tag summary display.

The one-line displays can be broken down into roughly two groups:

- One-line SCADA oriented displays
- One-line engineering or operational-planning oriented displays.

The one-line SCADA oriented displays are used for real-time monitoring and control of status (i.e. breaker, tap position, alarms) and analogue (i.e. amperes, watts). They are also used for displaying results from such programs as state estimation, short circuit calculation contingency analysis running in real-time or in extended real-time.

The one-line engineering or operational planning oriented displays are used by engineers and other users for input and detailed output from various application programs (i.e. optimal power flow, economic dispatch). These displays can also be used by the dispatchers.

Overview one-line diagrams show major transmission lines, interchange circuits and substations. Transmission lines and interchange circuits may be shown with total flow values, and indicators which show if energised. Substation one-line diagram and tabular displays show detailed information on the structure of the substation and current status of equipment and lines.

With respect to alarm displays a priority alarm list should be maintained. Alarms are added to the list in chronological order (most recent at the top) within each priority along with the category to which they belong.

The alarms can be in the following order:

- Unacknowledged major alarms
- Unacknowledged minor alarms
- Acknowledged major alarms
- Acknowledged minor alarms.

Each alarm message should contain the following information:

- Time of occurrence
- Station designation
- Point description
- Alarm condition
- Values and units.

5.6.4 Features, design, requirements

In this section the main features of an MMI subsystem will be discussed. The man-machine interface supports all other subsystems; concentrating all the man-machine communication into one subsystem results in a consistent, coherent interface concept for all functions. From the supporting role of MMI a very important requirement is emphasised concerning the overall software development that the MMI and CRT displays should be available at the earliest possible phase to provide a means of analysing the I/O of all applications.

The operator's interface must be simple to operate. SCADA/EMS experience is full of reports covering dispatch consoles with too many control switches or with overcrowded displays.

Most of the one-line diagrams are SCADA oriented and structured around individual substations. Not enough effort has been undertaken for a proper MMI design considering the specific needs of the advanced power application functions and those of the operational planning staff. The MMI has to be designed for both operators and operational planning staff's requirements.

The response of the MMI subsystem must be adequate to the operator's needs. Display of operator keyboard entries should appear within 0.2 seconds. Response within blocks of activity or sequences of interaction between operator and system such as calling a menu display, logs, etc. should require less than 2 seconds. Longer times are acceptable for computer processing of complex functions such as contingency analysis or economic dispatch.

It must be pointed out that full graphics is CPU and memory intensive. For a 350 nodes electrical network, the memory requirements could attain 20 MB.

The MMI design should also include innovative features such as 3-D representations, window technique, intelligent use of colours and shading capability. Geometric representation, such as circular profile (Kiviatgraph), takes advantage of the human visual system's ability to compare geometric forms rapidly. As the user becomes experienced with the circular profile, the asymmetrical polygons that result from off-normal situations will be easily detected as perturbations. A circle is created as the idealised system condition; this circle connects the end-points of many vectors of unit length if the value represented by each vector is normal.

The following conditions could be candidates for a geometric vector representation:

- System voltage
- Line loading
- Contingency line loading
- Spinning reserve
- Tie-line loading.

The MMI should use the advanced full graphics technology with the following features:

- PAN: to move around the electrical network diagram

- ZOOM: to concentrate to the level of detail required

- DECLUTTER: eliminate detail and non-relevant information

- CONTROL: control of image at dot/pixel level.

The full graphics technology permits:

- A better presentation of information
- An easier visual pattern recognition (e.g. the display of the line width could be made proportional to the ratio of actual load to desired/limit load)
- Supplementary features such as windows, viewpoints, superimposed data.

5.7 Inter-utility communication subsystem

The inter-utility communication subsystem is a software package that allows electric utilities with different suppliers' computer systems to exchange information.

The IUCS follows the guidelines established by the International Standard Organisation (ISO). The ISO defines a seven layer service architecture, with the lower three layers representing the X.25 protocol levels and the upper four layers representing communication service levels.

Demands for data from other utilities are constantly increasing as the task of effectively and economically controlling an interconnected power system continues to become more complex. Key data interchange areas include:

- Hourly schedules, actual interchange, generation outputs
- Power brokerage information
- System security data
- Shared generation participant allocation.

A UNIPEDE report reviews data exchanges between UCPTE, UFIPTE, SUDEL, NORDEL, CMEA countries and between systems in Great Britain. It shows the necessity of extensive data exchange for operational planning, daily operation and retrospective surveys. It recommends that measurements and indication data should be exchanged on a real-time basis for neighbouring networks contiguous to the national network and in the form of network equivalents for parts of the neighbouring system with less influence.

In conclusion, the benefits of national and international energy cooperation can only be achieved if there is some kind of information exchange between the partners of the interconnected system. This information exchange, mainly concerning network security, has to be intensified if benefits of interconnected operation are to increase. It is expected that this will lead to direct connections between computers in control centres of different utilities.

6 Power system control centre: dispatcher's activities

6.1 Introduction

In a SCADA/EMS control centre the operator is as much an integral part of the control system as the hardware and the software. More than that, the operator represents the most essential part. The operating and the operational planning personnel are the pivot around which the whole control centre revolves. The operator remains the ultimate decision maker because his responsibility for the power system is not separable. Thus the problem of an energy control centre can be approached as a system with three components:

- Hardware
- Software
- Personnel.

The efficient functioning of a control centre requires more than matching or juxtaposing these three sub-systems; it represents a synergistic effect, based on an appropriate combination of these components, where the human sub-system - the operating personnel and the operational staff - including the operating knowledge, philosophy, rules, and procedures of the electric utility constitutes the key.

As more and more automation is incorporated into a network control centre, the essential question becomes: what is a control centre or a control room, and what is the role of the operator within a control room?

The control room in a dispatch centre forms the human-operator interface to the process of power generation, transmission and distribution. This interface contains up to several thousands gauges, meters, digital read-outs, buzzers, flashing lights, diagrams, displays, etc. The control centre and the control room constitute the operational focal point and the nerve centre of the electric power system.

6.2 Salient features of the operator activity

Analysing the activities of a power system operator, it can be pointed out that in a steady, quiescent and normal situation he is working at a low stress level. This can change very rapidly when an emergency situation occurs.

The alternation between quiescent, boring, routine activities under low stress level and emergency, dynamic, non-predictable situations and actions under time pressure and high stress is characteristic of the activity of the power system operator. The decision-making process in a power system control room is realised through the human operators; as human problem solvers they are not alone. They are using and interact with a number of programs problem solvers or other intellectual tools.

One of these is the decision support system. The decision support system as part of the EMS is a collection of programs that constitute a sort of "crystal ball" for the dispatcher. It serves as a computer-aided consultant with advisory functions. The decision support system is divided into:

- Executive decision support system containing programs such as topology, dynamic network colouring, reactive power flow monitoring.
- Predictive/prophylactic decision support system containing programs for security assessment, economical dispatch.

6.3 A conceptual model of the dispatcher's activity

Fig. 6.3-1 describes a conceptual model of the dispatcher's activity. This model represents a cyclic process containing the following stages:

- Perception: obtain the current information
- Assessment: appraise the information by comparing expectations or patterns and identifying the significant features
- Planning : validate or revise current plans with respect to the most recent information
- Deciding : make decisions and select actions with regard to plans
- Execution : execute the selected actions.

With respect to the operator as a "problem-solver" three levels can be distinguished:

- The skill-based level
- The rule-based level
- The knowledge-based level.

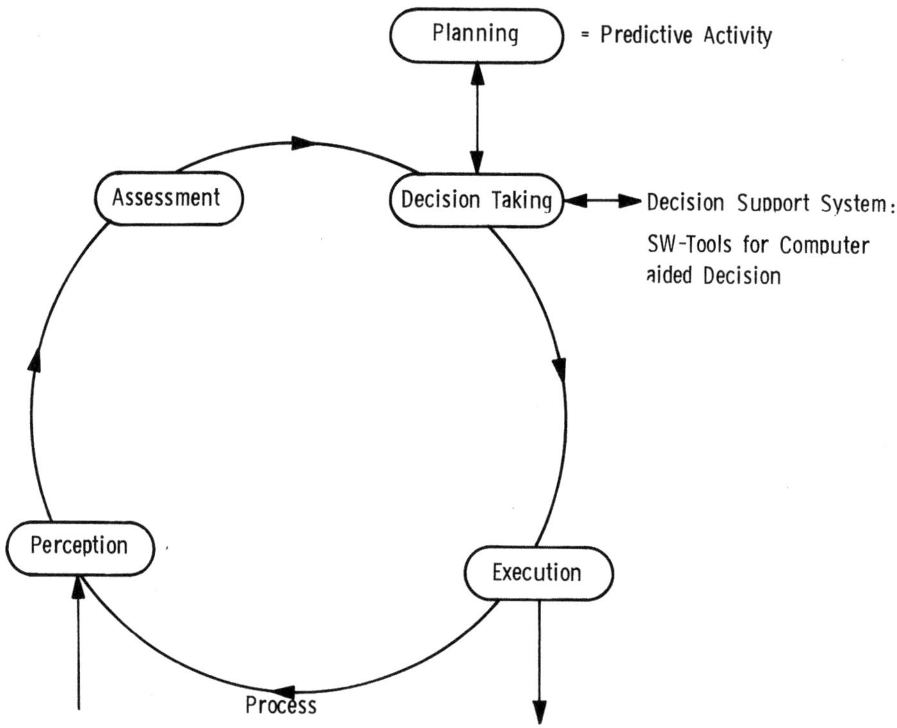

Fig. 6.3-1: A conceptual model of the dispatcher's activity

At the skill-based level, the operator must possess the ability to manipulate the devices available in the control centre. At the rule-based level, the operator performs a function and takes a decision when the incoming data flow matches a pre-defined pattern or solution. The rule-based level corresponds to situations where the dispatcher is forced to react to a given situation. At this level, situations may be learned from experience; that is a goal-oriented activity. In an energy control system centre the SCADA level corresponds at the operator's decision level with the rule-based level. At the knowledge-based level, goals based on an analysis of the power system are explicitly formulated. This type of knowledge is paramount to asking "what - if" questions. The executive and predictive decision support systems correspond to the knowledge-based level.

A very important aspect of dispatch activity is planning and personnel preparation to handle normal contingency and abnormal situations. In an energy control system, the ability to anticipate security and economy problems requires information from the EMS. A package of prophylactic/predictive functions offers the operator the ability to anticipate and to plan a logical sequence of events.

Equally important for the dispatcher is the "what - if" approach not only for the normal state but also for emergency situations. The main specific objectives and tasks of the operator include:

- To supply power to consumers without interruptions
- To minimise production costs, to save and optimise on interchange transactions
- To ensure the safety of field personnel
- To assure a coordinated maintenance program.

The Table 6.3-1 summarises the most important features of the operator's activity.

Table 6.3-1: Salient features of the operator's activity

Concurrent decisions

The operator must be able to handle events that occur at random times; he has to manage parallel activities such as data interpretation, short term planning, decision making, monitoring, commands, control.

Conflicting objectives

The operator must keep a good balance between conflicting objectives and requirements such as security and economy, timeliness and safety in switching and alarm clearance.

Decision making under uncertainty

The operator must be able to consider events which may or may not occur; he must take decisions based on the prediction of the future. He must cope with ill-defined or fuzzy situations. He must be able to make decisions with incomplete data.

Decision under overload and stress

The operator must be able to time-lag events and plan actions with knowledge of how long they will take. Operators often suffer from information overload, particularly during emergency situations. He must be able to make timely decisions under information overload, time constraints and stress; and to concentrate and focus his attention on the most critical problems first.

Heuristic oriented

The operator's activity is knowledge intensive. He learns best from experience. He is a heuristic and not an algorithmic oriented person.

6.4 Requirements

The operator must possess a thorough knowledge of the power system's characteristics, understand economics, communicate well, be able to use the computer resources and analyse carefully, but act quickly and decisively, and survive repetitive work and rotating shifts. It must be stressed that:

- Operator's mistakes are costly and may be visible while his saving or the interruptions he avoids could go unnoticed
- Power operators are not born with the necessary qualities
- There are no college/university programs to train power system operators.

Therefore much and continuous training is essential and necessary.

The following represent some desirable characteristics needed for power dispatch operators:

- Above average intelligence
- Multicausal thinking and the ability to make causal connections; analytical ability
- Utility experience and technical background
- Good memory
- Capability of being selective

- Capacity to anticipate phenomena, problems, critical situations
- Attention, ability to concentrate and to set priorities
- Flexibility
- Stress endurance, capacity to withstand being overloaded (stress carrying capacity)
- Self-confidence, self-assurance, capacity to inspire confidence and to exercise authority
- Determination, resolution, initiative and ability to take decisions on one's own decisiveness; ability to take responsibility
- Calm, "non-hyper" type; the operator should cope with "hours of boredom" and "seconds of sheer terror" - as in a black-out.
- Dedication, motivation
- Good health.

6.5 Trends in power dispatch operator's activity

Though power system functioning and operation are substantially more secure and economically, as well as qualitatively, better than years ago, blackouts and browndowns do still happen, often with dramatic consequences. It is an open question how far improved automation and computer penetration have contributed to the improvement in the operation of the power systems. Furthermore it remains to be answered if all the sophisticated algorithms and methods, the huge amount of data and displays, the very fast rate of evolving hardware generations is only an improvement for the dispatch of power.

Some blackouts in power systems, or the dramatic events in the nuclear power industry and the experience in the aircraft transportation industry shows that despite all the technical achievements the operator remains the principal cause of accidents and misoperations.

The Three Mile Island nuclear power plant accident and the Chernobyl nuclear-plant catastrophe are the best known recent events that have been attributed to human error. One conclusion may be that if humans are the major cause of black-outs, browndowns, energy interruptions, damages or even personnel casualties, then removing them from the control loop should produce an improvement in the power system operation.

If perfectly competent, reliable automation were immediately possible, this might well make sense. This is not the case. On the other hand a great number of unexpected events and situations were avoided because the operators reacted skilfully and rapidly; such events could not, at this stage of power dispatch technology, be even approached by the existing hardware and software.

Analysis of information in such cases has identified both "behavioural" and "system" factors associated with these events. It could be assessed that a computer or an expert system performing the same task as the power dispatch operator would have been less susceptible to behavioural factors (less sensitive) and more likely to ignore the system-related and heuristic factors.

Computers do not become complacent or yield to distraction; they remember without error and follow standard patterns. On the other hand, within the limits of today's computer technology, human operators maintain the advantage in reacting to, or mitigating some of, the system factors.

Computers cannot yet surpass humans in situations where information is lacking; degraded information or ambiguous conditions can create particular problems for machines. Humans have the ability to cope with ambiguous, vague or uncertain environments. They can make inductive decisions in new situations and generalise from previous experience. They can improve and exercise their judgement. Finally, though developers of knowledge-based systems are making rapid progress, one is aware that the knowledge-base represented by a power dispatch operator with thousands of hours control centre experience is of orders of magnitude beyond anything conceivable compared with today's level of technology.

The power dispatch operator has to remain a critical part of the control centre in the foreseeable future.

When SCADA, AGC, EMS, and the dispatch training simulator were developed and introduced between 1960 and early 1990, the overall reliability, economy and quality of electricity supply improved dramatically. The control centre design has benefited greatly from the computer revolution; the role of the power dispatch operator has also undergone a significant change.

Modern dispatchers have become more like managers than controllers or operators. They spend an increasing proportion of their time monitoring, retrieving and updating information, analysing trends, and using computer programs. Most of the power dispatch engineers are pleased with their higher technical roles and are using the new systems. But they are also concerned about the possible "erosion" of their manual operating skills especially in emergency cases. The solution is more extensive training and new forms of training that can keep pace with the increased complexity of control centre technology.

Another kind of operator error is emerging. Errors are being made in interactions with automatic and computer systems. Thus on the one hand it is clear that modern, highly computerised control centres are more reliable and efficient and represent a higher technological level than the old control centres supplied with telephones, analogue instruments and a mimic board. On the other hand, the problem of human error has not yet been solved. The human error of the power dispatch operators has changed in cause, kind and consequences.

Another matter of great importance is the increasing significance of developing a comprehensive model of operator behaviour. Despite all the impressive achievements in control centre technology it must be noted that the computer has replaced only the classical operator's monitoring and controlling functions. They are algorithmic and for the most part based on well posed problems of power system analysis and control. In a sense they can only respond to situations anticipated by the power dispatch operators.

By contrast, the basis of the new approach to power system control centre automation encouraged by the increasing field of artificial intelligence, has been the implementation of expert systems. This represents a revolutionary advance in the evolution of power system control centre technology (see chapter 10).

Expert system applications must be carefully selected, designed and integrated into the control centre based on a full understanding of the operator's tasks, requirements, salient features and capabilities.

Personnel in advanced power application functions aim to serve competently and efficiently, i.e. like a consultant engineer. Like a pilot, a power dispatch operator is engaged in the moment-to-moment control of the power system and is responsible for making virtually all real-time decisions. Advanced power application functions are mostly of a preventive/predictive nature.

Other functions, like alarm handling must be active during incidents and emergencies; in such cases so many alarms are triggered simultaneously that the power dispatch controller is confronted with a bewildering barrage of noise and signal lights. Thus, a qualitative new function, i.e. alarm processing, will be necessary.

In conclusion, the philosophy concerning the role of the power dispatch operator in the future can be described by the following short formula: keep the operator in the loop but give him the intellectual tools permitting him to stay ahead of the power system operation.

7 Power system and dispatch training simulator

7.1 Introduction

Complex technical processes require careful training of operating personnel. This is particularly true for systems where training might affect equipment or safety of people if done during regular operation. Important applications are flight simulators and power plant simulators. Both have reached a very high level of maturity and their acceptance is not questioned any more.

The process control computer between the power system and the operating personnel has two effects:

- It offers a great capability for systematic information processing with highly developed software tools.

- The occurrence and the effects of very large disruptions have been reduced due to the implementation of modern SCADA/EMS.

Operating personnel are able to manage the complex power system during normal conditions in a most secure and economic manner. However, during rare but serious emergency situations, human reactions are not always correct because it is not possible to familiarise the operating personnel with all features occurring during emergencies.

The main requirements for developing and introducing a power system training simulator for the operating personnel are the following:

- Develop a collection of scenarios for the operating situations most often encountered and for possible emergency situations through system analysis.

- Invest the resources in manpower to install, maintain and enhance the presently available and newly introduced energy management functions for normal and emergency situations.

- Sufficient real-time data and system knowledge must be made available through systematic data collection.

- The development of computer-based tools for inter-control centre data exchange, for the fast and accurate detection of vulnerable system conditions and for the user-orientated data display is urgently required.

There is still a gap between the conceptual capabilities of modern energy management functions and their practical effects for system operation. A power system training simulator will help to bridge this gap.

As a by-product the training simulator may constitute an important step towards an expert system. It may contribute towards the construction of a process knowledge basis that could be used by an expert in order to extract operational strategies for normal and emergency situations. The simulator will serve as a tool for the shell expert in order to extract the process knowledge from the dispatcher, i.e. the simulator allows a better insight and understanding of the operator's behaviour.

7.2 Power system technological requirements

The generation, transmission and distribution of electric energy is characterised by a very high degree of reliability, i.e. supply interruptions occur very seldom. It is almost impossible for the operating personnel to be trained on the job for emergency situations. Furthermore, the disruption pattern for possible emergencies is very large and hence the repetition of a specific emergency situation can almost be excluded.

The technological innovation rate for the monitoring and controlling equipment is very high. The possibilities offered by modern data processing systems have increased considerably during the last ten years. Hence, new software tools have been developed and implemented to support decision making by the operating staff. One purpose of a training simulator is to familiarise the operators with new methods and to develop the necessary experience as to the situations they may encounter.

Operating personnel of generation and transmission systems have to meet functional requirements quite different from those encountered in a distribution system. Hence, in the former case, training is mainly required for such areas as economic load dispatch, transmission scheduling, stability problems, emergency control, etc., whilst in the latter training is needed for switching operations under normal and emergency conditions. A training simulator must consider the specific requirements of the utility. In view of the large number of existing companies the need for a wide range of solutions is obvious.

Control centres are tailored to the specific requirements of a power utility. As the training simulator must be based on realistic power system and control system models, each company needs its own specific training simulator. This point must be emphasised because it is an important cost factor in the implementation of a training simulator. Unlike the aircraft industry where a plane is manufactured for many different airline companies with essentially the same cockpit and consequently the same flight simulator, each utility puts a lot of effort into the specification of an individual control centre optimally suited to meet its operational requirements. As a consequence, the degree of duplication from one training simulator to another is limited.

Finally, it should be mentioned that the qualification requirements for operating personnel are very different. These range from high school diploma personnel with practical experience to the university career engineer. Hence, the motivation and necessity to introduce training simulators is very different from one utility to another.

A power system training simulator is necessary to enhance the capability of the operator in order to achieve a safe, secure and economic operation under a large variety of operating conditions meeting all demands with respect to the quality of electric energy. The aim is to increase the operator's knowledge of the:

- Power system behaviour under normal and emergency conditions
- Reliable and fast interpretation of the relevant system information
- Available energy management functions for system monitoring and control.

Fig. 7.2-1 shows in a schematic manner the information flow in a power system training simulator. The simulator consists of two models: the power system model and the control centre model and two man-machine interfaces, one for the trainee and one for the instructor. The information exchange between the power system model and the control centre model with the trainee's console occurs as in the real system. The instructor has both the functions of scenario creation and training management. This aspect is further explained in

Fig. 7.2-2 by the management outline of the instructor and the control flow of the trainee. It is obvious that all SCADA and EMS functions must be available to the trainee in order for him to have working conditions identical to those in the real system.

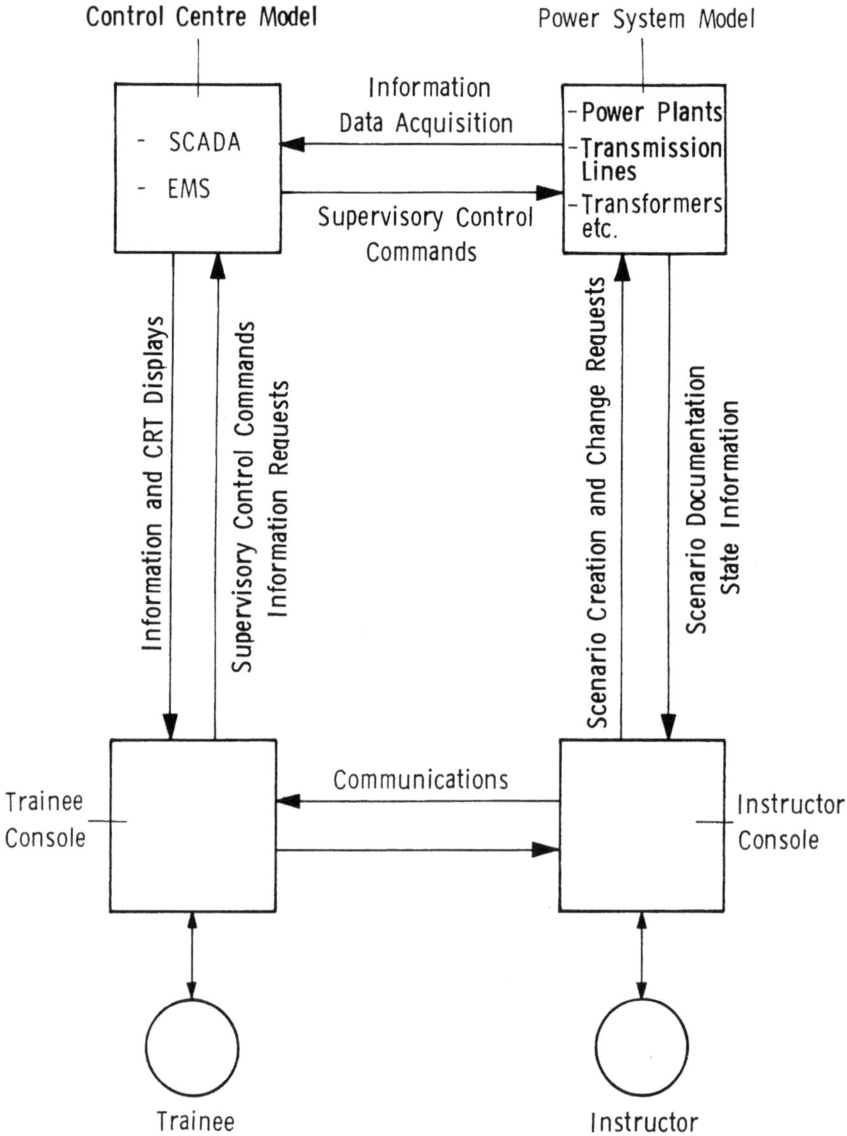

Fig. 7.2-1: Information flow in power system training simulator

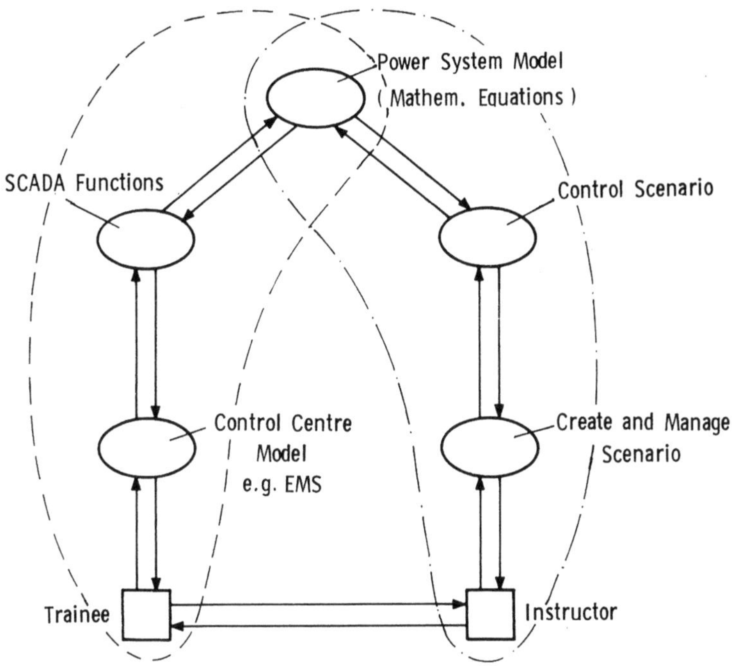

Fig. 7.2-2: Scenario management and control flow

A training simulator may also be very useful to different groups of users as listed in Table 7.2-1.

Table 7.2-1: Different groups of users for a power system training simulator

Group of users	Aim/use
Operators/dispatchers	- Basic operation - SCADA functions/switching - EMS functions - Emergency control - Restorative control
Operation support staff	- Short-term planning - Post-event/post-mortem analysis

	- Development and validation of new operational procedures
Software support personnel	- Software maintenance - Trouble shooting (debugging) - Software-enhancement/development - Test bed for EMS program development and checkout
Hardware support personnel	- Hardware-maintenance - Trouble shooting - Hardware-enhancement
Consultant and suppliers of control centre	- Design, development and testing of new control centre systems
Power utilities	- Test of criteria and allow to demonstrate the capabilities of the newly acquired control centre - Cost reduction for personnel training - Time reduction for implementation

7.3 Functions of a training simulator

7.3.1 General functions

The primary purpose of a training simulator is to provide a realistic environment in which system control engineers can be trained. The facilities provided and arrangements for access must recognise this main objective. These facilities will, however, be similar and superior in some aspects to those normally provided for in operational planning and extended real-time studies.

There are two broad types of training simulator:

- Generic, in which performance of the operational system is modelled with no attempt to replicate the actual system or the man-machine interface. An electronic analogue analyser could be considered an early form of a generic simulator.

- Replica, in which the performance of the actual system is modelled and actual or close approximation to the operational man-machine interface is provided.

Turning to the forms of training required, several aspects can be identified:

- Training in the use of control room facilities such as SCADA and EMS, communications, documentation and procedures. It is most important that the control engineer should be familiar with the use of all the facilities in the control room. Although a simulator could be used, training of this type is probably most easily and effectively provided on the operational equipment.

- Training in switching procedures and safety related matters. Some control engineers, both at the control centre and other control locations, will need training in technical aspects and procedures for switching. This can be done conveniently on a training simulator, requiring a man-machine interface of the type used operationally with appropriate simulation of SCADA and substations. The instructor would act as the remote operator. It is essential that detailed substation and system displays used in the operational SCADA are available.

- Training for operation during normal conditions. The training required is essentially in the technical aspects of system operation during normal and near normal conditions. It may include the following tasks:

  - Load-frequency control
  - Economic dispatch energy exchange
  - Monitoring of system conditions
  - Use of the contingency analysis and operational load flow facilities
  - Interpretation of alarm displays
  - Switching
  - Load management.

  Many of the existing training simulators, whether stand-alone or integrated with the SCADA and EMS, provide this form of training. The simulator may be extended to model "simple" fault conditions, i.e. those for which all generators can be assumed to remain at the same relative phase and static power flow models are adequate. To provide all the facilities listed means, of course, that the simulator must have access to many of the computational aids available to the control staff in real-time operation.

- Training for operation during emergency conditions. A real-time dynamic simulator is the most effective mechanism to train control engineers in the handling of severe disruptions. The broad objective of this training would be:

  - To increase confidence in ability under stress, to weigh up situations and make timely decisions.

  - To improve knowledge of the technical characteristics of the system under dynamic or degraded operating conditions.

  - To improve knowledge of procedures and facilities for handling emergency situations.

A replica simulator incorporating much or all of the operational man-machine interface, as well as a model of the actual power system dynamically reacting in real-time to internal and external signals, is probably the only way in which the degree of technical and physiological realism can be achieved for a control team. Ideally, it should be possible to model the following types of incidents:

- Multiple, simultaneous or sequential faults
- Protection operation (overcurrent, impedance, load shedding and other automatic switching schemes)
- Oscillatory conditions including non-uniform oscillations of different machines
- System splitting and islanding.

It is not difficult to carry out the relevant computations in the generous turn-around times of many seconds, minutes and even hours often acceptable for operational planning work. The essence of a training simulator, however is, that the system information should be presented on SCADA timescales, e.g. an update every few seconds. This adds substantially to the technical problems, in particular to dynamic and transient modelling.

7.3.2 Specific functions

In view of the modelling requirements, the main training tasks are grouped into the following four classes:

Class 1: SCADA functions

Training in this class include supervision and switching functions such as those listed below:

- Normal switching operations
- Voltage and reactive power control
- Limit value monitoring
- Failure of data transmission equipment
- Line or transformer breaker trip
- Generator trip.

Class 2: Steady state functions

Training functions of this class are related to the instantaneous loading of the system and power station control and comprise, amongst others, the following functions:

Security:
- Topology monitoring
- State estimation
- Operator load flow
- Contingency analysis
- Balanced and unbalanced fault simulation
- Reactions to faults, e.g. corrective switching.

Economy:
- Economic load dispatch
- Interchange scheduling
- Optimal power flow
- Load management.

Class 3: Dynamic functions

This class of training functions requires sophisticated modelling of the power system. Large computing power of the modelling equipment is therefore normally necessary. Functions such as those listed below are incorporated in this class:

- Load frequency control
- Voltage and reactive power control
- System split and system islanding
- Load shedding
- Energy control, e.g. system restoration.

Class 4: Energy management functions

These training functions are based on tasks for short-term (daily) and mid-term power station scheduling. Programs available in the control centre system for this purpose can be incorporated in the training simulator. The following functions are examples of these types of training tasks:

- Load forecasting
- Unit commitment
- Interchange evaluation
- Energy negotiations.

It is obvious that the complexity of the power system model depends to a large extent on the type of functions for which training is required. If quiescent behaviour is assumed, the modelling of the network for training of functions under class 1 can be based on normal load flow calculations. SCADA programs and programs for steady-state security assessment available in the control centre system can partly be used in the training simulator. In a similar manner, programs available for load frequency control functions and - especially those for operation planning functions can be incorporated in the training simulator. The training for the more advanced functions under class 2 and 3 requires quite advanced network and power plant models as well as sophisticated simulation techniques.

7.4 Modelling aspects

A substantial part of the design of a power system training simulator is concerned with the development of a suitable power system model. The reaction of this model to the operator's commands should be as realistic as possible. Depending on the training functions, the model has to contain static and dynamic properties of the power system. The following general remarks determine the complexity of the simulator.

For training under normal operating conditions, the model must work as closely as possible to real-time. The capacity of the available computing equipment essentially determines the realism of the training system model. Power system dynamics may be modelled using the coherency assumption, i.e. the simulation can be limited to mid-term and long-term dynamics. They include the calculation of frequency deviation after loss of generation or load.

The voltage behaviour may be based on simplified generator models. The dynamics of the voltage regulator and excitation system need normally not be included and are approximated by step changes. Modelling of the loads should include realistic voltage and frequency dependence. As far as real-time requirements are concerned, the simulator model should allow realistic training for the following tasks:

- SCADA functions, e.g. switchgear operation, voltage control, effects of protection relays operation, etc.
- Automatic generator control functions such as load frequency control, load shedding etc.

The complexity of the simulation model is limited by the following two constraints. The model must not be based on parameters which are not available in practice because of the use of unreliable parameter values may lead to a model behaviour not being typical for the actual system. The second constraint is the computing speed. The more complex the model - especially the dynamic part - the more computation time is required. Use of parallel processor equipment may improve the speed of simulation.

A training simulator may be of particular use for the study and analysis of specific emergency situations. In this application, the aspect of real-time simulation may be of secondary importance. Here, correct representation of the power system, including aspects of data transfer to the control centre and of the facilities offered by the control centre system define the predominant requirements for the simulator. Particularly in countries where the operating personnel have a high professional education level, emergency training is an important justification for the use of such equipment. Lack of operator experience during emergency conditions may contribute substantially to handle major disruptions.

At this stage emphasis should be made regarding modelling and the data problem. In order to have realistic operating conditions, the training simulator should be linked to the system data base. The possibility of updating the model with actual system data is an important prerequisite for the efficient use of the training simulator.

Fig. 7.4-1 shows the simplified structure of a system model and the classes of training functions discussed in section 7.2. The functions for which training is required determine the complexity of the system model. According to the classes of training functions, the following four levels of system model complexity can be defined:

- For the training of SCADA functions (Class 1), a steady-state power system model is normally adequate. It is based on the static load flow equations and may include the load models and effects of protection relays operation (overload, underfrequency, etc.)

- For the training of the load frequency control (Class 3) the model must contain at least simplified power plant dynamics.

- In order to simulate complex emergency situations (Class 3) including stability problems, modelling and data problems increase considerably due to the need for a more detailed consideration of the system dynamics. This requires highly developed computing equipment.

- The model for the training of energy management functions is mainly based on load curves, power plant data, contracts for import/export of electric energy, etc. For network losses and restrictions in energy transmission, additional modelling of network elements may be necessary (e.g. optimal active power flow). Programs available for solving operations planning functions in the control centre system can be used for training in this field.

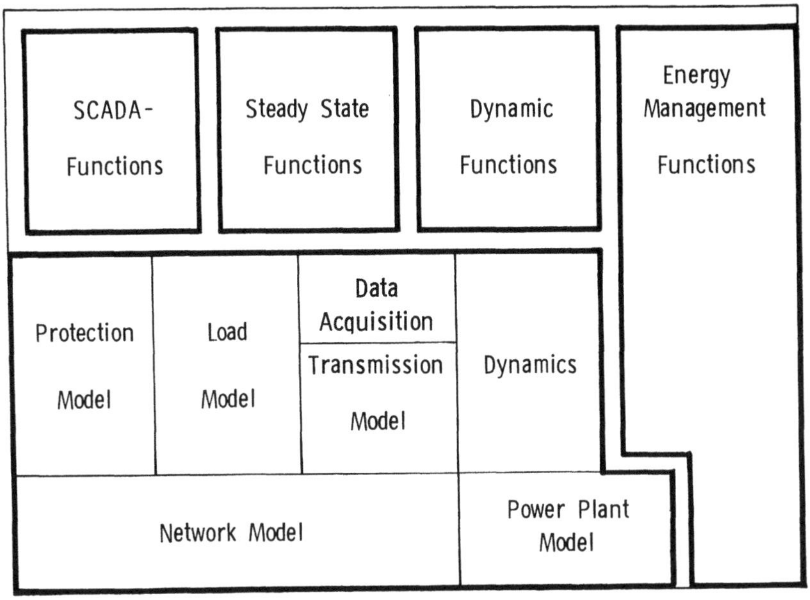

Fig. 7.4-1: Simplified structure of the training functions of the system models

It should be emphasised that the development of a real-time simulator for conditions ranging from normal to emergency requires a suitable dynamic model. Furthermore, it must be implemented in an interactive manner, so that trigger conditions from the trainee, the instructor and the model itself can be implemented without any restrictions. As the available computer power is limited a very careful compromise between accuracy and realism of the model has to be found. As the SCADA system provides information of the operating conditions to the trainee fast transient evolutions of the system state are not known to the operator. Two types of phenomena have to be distinguished:

- Mid-term phenomena directly observable in the control centre model including voltage collapse problems.

- Short-term phenomena not observable in the control centre model. However, such phenomena may have very important consequences on the system state.

The latter are calculated in an approximate manner using a simplified transient stability program. It calculates the immediate reaction of a highly disturbed system and deduces from it the final state on a simplified criteria basis.

As far as the mid-term dynamic model is concerned the following properties are important:

- The static load flow model contains all relevant elements (lines, transformers, etc.).
- The loads are frequency and voltage dependent.
- The generating units contain models for the boiler, turbine, speed regulation, load-frequency control and voltage regulator.
- The protection devices are modelled with respect to overload, load shedding, low and high frequency, under voltage, short-circuit protection and loss of synchronism.
- The network topology defined by the switching devices, busbars, etc. is fully displayed to the operator, i.e. he may address individual elements of the power system through the available telecontrol model.

Fig. 7.4-2 shows one time step of the dynamic power system simulator. The electric state (node-branch model) is calculated from the element information describing the topological state. This step requires a very large data flow. This is typical for a power system training simulator in contrast to a general power system simulator which is directly based on a node-element model. The generation unit states describe the initial conditions of all units at the beginning of the time interval (T, T + h). Once the electrical state is determined (which requires n iterations of the load flow equations) the control commands by the trainee and the trainer, the possibly pre-programmed events and the operation of the protection devices are all taken into consideration. The possible occurrence of a serious emergency is checked within each time interval.

If it occurs, a simplified short-term dynamic simulation is started in order to assess the effects on the power system state. Finally the mid-term dynamics calculate the system state at the end of the time interval T+h. The time step length h is determined by the cycle time of the SCADA system and lies in the order of 2 to 10 seconds.

The long-term changes are entered by the instructor as a series of events either at the beginning or during a training session.

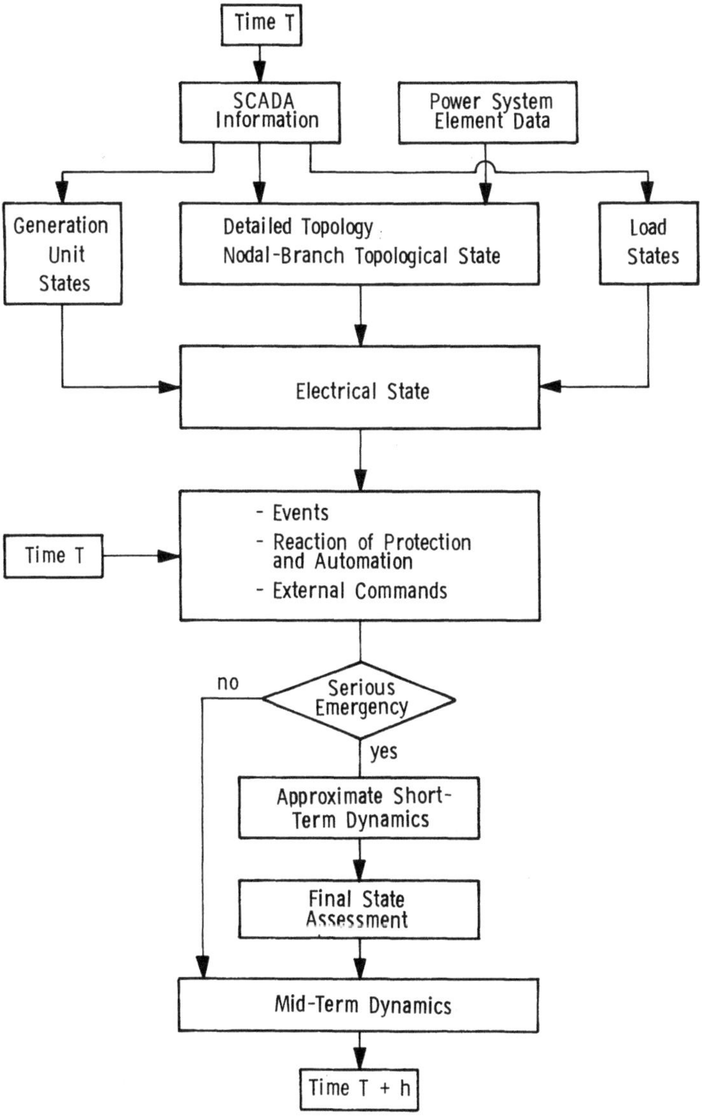

Fig. 7.4-2: One time step of the dynamic dispatch training simulator

7.5 Different types of training simulators

7.5.1 Stand-alone version

This version is shown in Fig. 7.5-1 and its attributes can be defined as follows:

- Normally no direct connection to the control centre system exists.
- The hardware/software is in most cases different from the hardware/software in the control centre system.
- The man-machine interface (MMI) is normally not identical to the MMI in the control room.

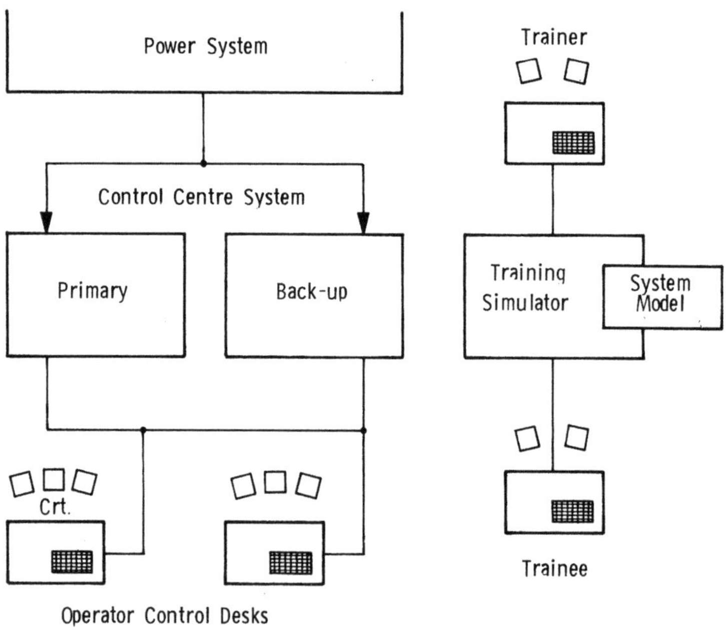

Fig. 7.5-1: Training simulator in a control centre as a stand-alone version

Features of this training simulator version are:

- It is independent of the functions implemented in the control centre system and therefore suitable as an initial step for training independent of the hardware/software of the control centre system.
- It is flexible as regards the place and time of installation.

- Basic systems are available on the market.
- The common use of a power system training simulator with other companies is possible.
- Adaptations of the basic system to individual needs especially in the field of man-machine-interface and operations planning are necessary for realistic training.
- The procedure for filling in and amending system data (description of switching and alarm elements, measurements, topology, characteristic data of lines and transforms, etc. as well as the construction of CRT-graphics and forms) is normally different from the inputting of this information to the control centre system. Thus, normally, this information has to be entered twice.
- There are limited possibilities for training in a man-machine environment identical to those in the control room.

7.5.2 Integrated version

This version of a training simulator is shown in Fig. 7.5-2 and its attributes can be defined as follows:

- It is linked to or integrated in the control centre system.

- It uses the same kind of hardware as the control centre system (e.g. it runs on the back-up computer or on an additional computer installed for training, software development which is sometimes applicable as second back-up). Where computing capability has been exceeded, e.g. when modelling the dynamics of the network, the power system model can be installed on separate hardware interfaced to the training computer (indicated by dotted lines of the system model block in Fig. 7.5-2).

- It utilises the main parts of the control centre system software (SCADA, MMI, storing, load flow, functions of operating planning, etc.) and control centre data (power system descriptions, graphics, forms, network image, actual and archived voltages, injections and loads etc.).

- The MMI is of the same type as that in the control room except for the mimic diagram (if it exists).

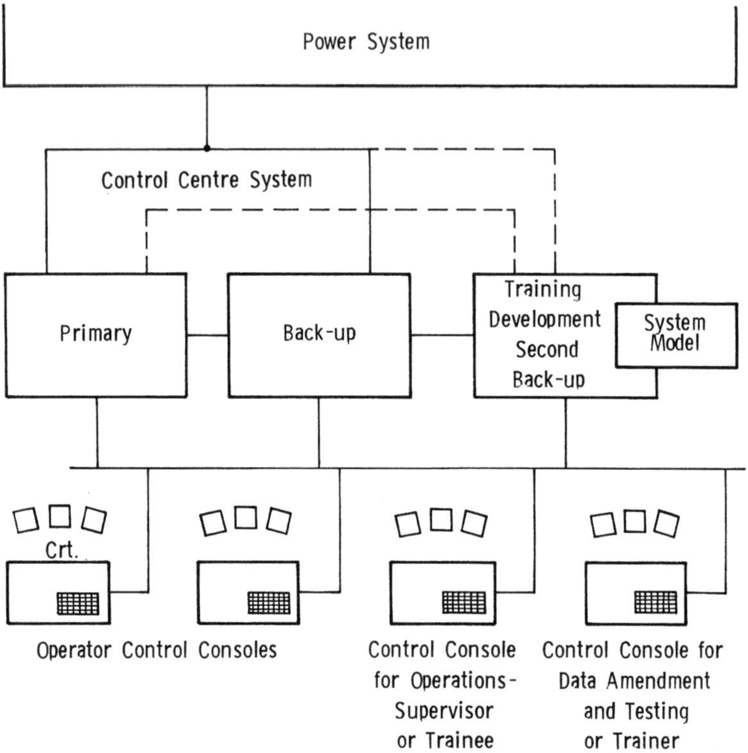

Fig. 7.5-2: Training simulator integrated in a control centre system on a third main computer

Features of this training simulator version include:

- The basic functions used for training are normally the same as those used in normal operation in the control room.
- Most system data, graphics, forms, etc. are transferred from the control centre system, thus avoiding separate filling in and amendment of these data.
- In some projects, control consoles that already exist in the control centre system (e.g. the control console of the supervisor in the control room and the control console of the operator in the computer room) can be utilised for training purposes. In this way, hardware can be saved.

- Some manufacturers have included software for training in their control centre software packages.

- Adaptations of the control centre software and of the standard software packages for training are necessary to cope with the needs in the individual project.

- Training functions can be used to support testing and the factor acceptance test of a new control centre system.

- Operator training on a new control centre system is improved when training functions are incorporated into the system.

7.5.3 Implementation

Stand-alone simulators were first implemented using hybrid computers. The power system is modelled with a special purpose analogue computer while its operation and control is implemented through use of a digital computer. The main disadvantage of this implementation is the lack of the power system representation on the element level. Digital simulators for stand-alone versions have the same drawback and may only serve as a training tool for basic power system behaviour. Furthermore, additional expenses may be caused by the needed hardware. It becomes difficult and expensive to build such simulators for an established control centre; some functionality of the on-line EMS is lost, such as the use of real-time data as the basis for training. Effort is duplicated in maintaining two separate data bases.

The advantage of the integrated training simulator is that it can be implemented more economically if procured along with the control centre since it may use the back-up computer. It offers additional advantages of a unified data base and the ability to retain real-time data for use in a training exercise.

The disadvantages of using the integrated training simulator on the back-up computer are that additional complex software is required to ensure isolation of the simulator from the on-line system. If the primary system goes down, failover occurs and the training session is terminated or there may be competition for the use of the back-up computer system for software development and training.

The use of a second back-up computer for the implementation of the power system model linked with an interface to the control centre model available to the trainee is therefore recommended.

7.5.4 Economic considerations

The qualitative economic justifications for the implementation of a training simulator are:

- Shorter time for operator education and training
- Full and effective use of the available software in the control centre through the EMS
- Improved power system economy and reliability under normal and disturbed system conditions
- Reduction of the system restoration time after long disruptions.

As mentioned before, the necessity of introducing a power system simulator depends very much on the professional qualifications of the operating personnel. The higher the qualification attained, the less is the necessity for training under normal conditions. Yet training under emergency conditions requires some kind of a simulator. The main problem towards the implementation of a training simulator is the cost of the required hardware and software. While the power system model is more general its interface to the control centre model is very specific and utility dependent. Furthermore, the interface between the data bases of the real system and the real-time simulator requires much control centre and supplier specific problem-solving. Experience with the data exchange using standardised protocols between different control centres may greatly affect the economic factor.

7.6 Training scenarios and training sessions

Based on the previous section on the topic of training scenario building, training session design will be restricted to the integrated version only. Fig. 7.6-1 shows a simplified training situation, both from the trainer's and trainee's control desk.

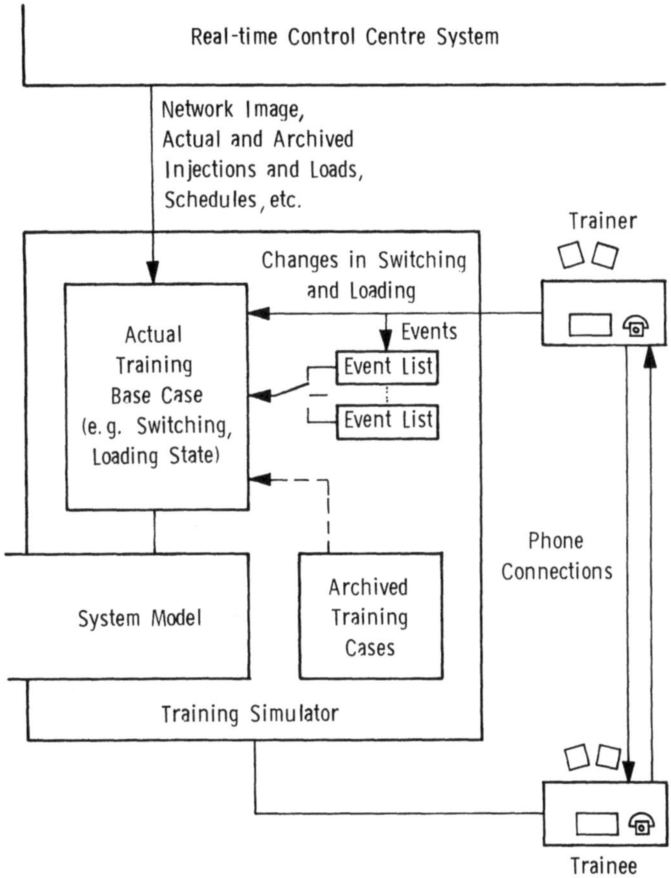

Fig. 7.6-1: Man-machine interface including examples of data entry possibilities for training

The control desk for the trainee as well as that of the trainer is of the same type as that used in the control room. In a training session, the trainee is in a situation that can be described as follows:

- He reacts as he would do for the task for which he is being trained: e.g. network supervision and control or power station operation planning.

- A situation is presented to him in the same way as to an operator in the control room.

- He has to react to the system situation and the system behaviour as well as to events introduced by the instructor.

- Additional advice can be given by the instructor via the phone connection.

- The trainee should address orders that are given in control room practice via the phone to the instructor in the same way and respond in the appropriate manner.

In addition to the functions that can be utilised by the trainee, the instructor has the following possibilities. He can establish a basic situation for a training session by copying data base areas such as network topology (switching state), network loading curves (injections and loads), power station and interchange schedules from the actual control centre system or by calling a basic situation buffered in the training system from a previous session. In addition, tools are supplied for modifying this information:

- He can select the time scale for the training program and the system response. The minimum time scale is given by the response time of the programs modelling the power system (e.g. the load flow calculation). A reduced time scale may be suitable for training in operations planning.

- Events such as failures in data transmission, earth faults, breaker trip, generator trip, etc. can be introduced into different training case lists defining events and events time related to the starting time of the training period. Depending on the reaction of the trainee to the events, the instructor is able to change from one training list to another, as well as to introduce new events via his control desk.

- The timer for activation of events in training case lists can be started and stopped via the instructor's control desk.

- Consecutive events can be defined by the instructor or can be generated by the model itself. For example, a breaker trip caused by a line or transformer overload persisting longer than a given time interval can be simulated in this way.

- When defining a training session and examining the reactions of the trainee, the trainer himself is in a training-like situation also learning how to improve his teaching methods.

It should be pointed out that the instructor is required to play many roles, e.g. he acts as generating station or substation personnel or high-load customer, etc. The development of realistic scenarios is a difficult task which requires a deep insight into power system behaviour. Special software tools are necessary for an event-oriented simulation. The sequence of events has to be adjusted to the trainee's capability and experience.

7.7 Concluding remarks

Since training is the key to success the power system and network training simulator is now used comprehensively by utilities throughout the world. As information processing becomes an integrated part of the power system operation, its effective use under a wide variety of operating conditions becomes mandatory. However, as actual operation is restricted to normal mode it is not possible to become familiar with realistic emergency situations. Increased complexity of the power system with narrow operating limits and high pressure toward economic operation make operator training necessary. At present, several utilities have decided to install stand-alone or integrated simulators: in Japan eight of the nine large utilities have training simulators available, and the American EPRI is sponsoring the development of a large training simulator. In Europe several utilities plan to install training simulators for their control centre staff.

In conclusion, power system training simulators become more important as the computer systems for monitoring and operating electric power systems increase in complexity. The implementation of the training simulator depends very much on the educational level of the operating personnel. Finally, the structure and the tasks of the utility have a great influence on the simulator design.

The present trend is towards an integrated version of the simulator in the computing equipment of the control centre system. The initial implementation of an integrated training simulator is not too costly as long as the available hardware and software tools on the on-line system can be used as a basis and provided the system dynamics need not to be modelled in detail.

Obviously, an initial version like this can be used in training for a limited number of operational tasks.

For the training of the full range of operational tasks, the complexity of the system model increases. In view of the real-time requirement of the simulator, it is necessary to implement the simulator on hardware other than that of the on-line system.

8 Existing energy management systems

8.1 General remarks

This chapter describes three types of modern control centres. It must be emphasised that the control centres in operation today throughout the world cover a wide range with regard to both hardware and software. Operating strategies as well as economical and technical constraints lead to a very great variety of control centres. It is interesting to analyse all these constraints in view of the different control concepts which have been implemented. A selection of three different control concepts from North America, Europe and a developing country may give some insight into the large spectrum of solutions already realised.

8.2 Energy management systems in a US utility

8.2.1 System characteristics

The utility considered has an installed generating capacity of about 22 000 MW, of which 90% is coal-fired, 8% is nuclear and the remainder is hydro, pumped storage hydro and combustion turbines. The transmission system includes about 3200 circuit km at 765 kV, 180 circuit km at 500 kV, 6000 circuit km at 345 kV and 13 200 circuit km at 138 kV. Reliable and economic operation of the transmission network is an important prerequisite for the overall bulk power supply reliability. Over the past years the utility has established more than 140 interconnections with 27 neighbouring utilities; hence it is an important partner within the interconnected network as it contributes substantially to the economic operation by:

- Providing mutual assistance when generation capacity deficiencies occurs;
- Selling excess energy to neighbouring systems to maximise the utilisation of its own generating capacity;
- Enhancing system reliability through coordinated planning and operation.

In order to make full use of these possibilities it is imperative that there is an exchange of operating information with the interconnected systems.

In developing the transmission network to assure economic and reliable operation, sufficient capability and flexibility has been provided in the network to accommodate a wide range of power flow patterns that occur in day-to-day operation and still have the ability to withstand severe emergency conditions without uncontrolled cascading trip-outs of system equipment. The transmission network must not be a constraint in the utilisation of generating resources and the system's ability to service its load requirements or exchange power with neighbouring systems.

8.2.2 The concept of hierarchical control

A three-level hierarchical organisation to carry out the operational control and security assessment functions has been developed. The main control (MC) centre dispatches and controls the generation of all the operating companies including:

- Economic dispatch
- Load frequency control
- Unit commitment
- Maintenance scheduling
- Power exchange with interconnected utilities.

The MC centre also monitors the operation of the bulk transmission system to assure safe and reliable operation under all conditions.

There are four regional control (RC) centres located in four major operating company headquarter cities. The RC centres are responsible for:

- Operation and coordination of their respective transmission systems
- All manual switching operations in their own systems.

Portions of their transmission dispatching functions are delegated to eleven geographically dispersed district control (DC) centres that provide transmission coordination.

Clearly, there is a need for overall coordination of the activities at the three levels in this hierarchy and this has evolved into a very close working arrangement among the various control centres.

8.2.3 Enhanced control and security assessment

Utilities in all industrial countries need to minimise future capital investments and maximise the utilisation of system facilities already in service. This leads to transmission networks operating closer to their design capabilities than was the practice in the past. Furthermore in the United States the effect of the large disparity that exists between the cost of coal-fired and nuclear generation relative to oil-fired generation has a great impact on the day-to-day operation and loading patterns of the transmission system. As a result of temporary surplus capacity substantial motivation exists to transfer economy energy from coal-fired regions to areas which are dependent on oil-fired generation.

To assure that the transmission system is maintained in a secure operating state, the system operating functions need to be reinforced by the addition of expanded capability to monitor the state of the system by more sophisticated and responsive means to assess the security of the system. Security assessment involves the ability to evaluate:

- The capability of the system to meet expected demands taking into account generation and transmission contingencies.
- The impact of operator decisions on the scheduling of generation, interchange, and transmission outages.
- The effect of strategies which the operators may use to alleviate potential system operating problems.

8.2.4 Implementation

An integrated enhanced security assessment and control system implemented at an MC centre is shown in Fig.8.2-1. The on-line MWh monitoring function comprises an information logging process which operates without interacting with other control or data acquisition functions except the system operator. Similarly, automatic generation control interacts automatically and independently directly with the generating units to control the output of each unit. The state estimation transmission monitoring function is the central part of the enhanced security assessment including external equivalents, contingency selection, load forecasting, on-line load-flow and corrective strategies.

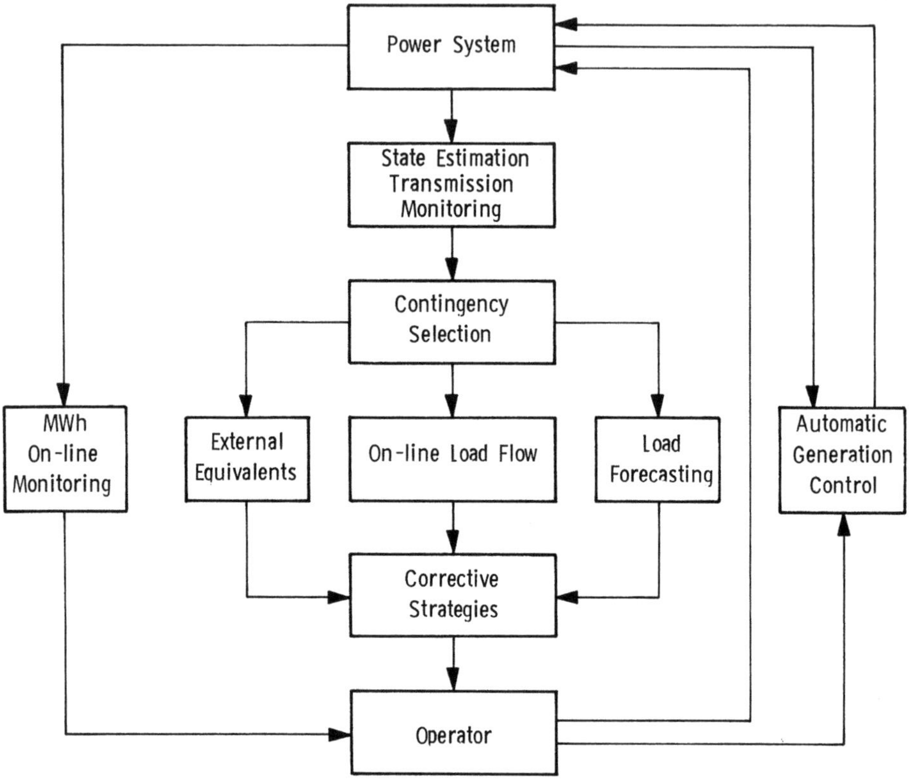

Fig. 8.2-1: Integrated security assessment and control system

Fig. 8.2-2 shows the arrangement of hardware within the control centre hierarchy and the communication links between them. In the MC centre two VAX 11/780 computers are installed with associated CRTs, line printers, dynamic system display board, and assorted graphic recorders. The equipment at the RC centres consists of two VAX 11/750 computers and associated loggers and CRTs. Some of the eleven DC centres are equipped with dual VAX 11/750 computers to provide facilities for the transmission line data acquisition. These computers act as data concentrators to transmit appropriate data to the next level.

The automatic generation (AGC) control function at each generating plant is based on remote terminal units (RTUs) directly connected to the unit controls. In addition to MW other quantities are transmitted to the MC centre including MVArs, auxiliary load, actual limit setpoints and the status of the control system.

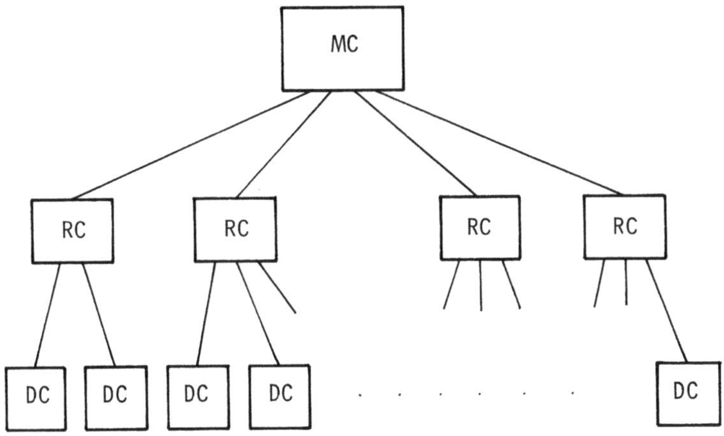

Fig. 8.2-2: System control computer configuration

For communication a digitally based system is adopted using appropriate data formats and communication techniques. In this manner a substantial saving may be realised. In addition, the digital system permits the accommodation of additional functions easily and at relatively low cost. The control centre is serviced by a digital system from all plants for the AGC and from each of the four RC centres for data acquisition and the interconnection. The digital system also includes communication between RC centres and the appropriate DC centres.

8.3 Energy management systems in Germany

8.3.1 Introduction

In Germany the generation of electricity is performed by public power supply, industry, and coal mining companies. Public electricity supply companies have the largest share with 85% of the total electric energy production reaching 418 TWh in 1987.

All companies which distribute electricity for supply to the consumers are public utilities. Because of the liberal economic system many companies of a very different capital structure have a share in the public electricity supply. In line with the importance of the supply area the prevailing number of the electricity supply companies are either publicly owned companies or public-private companies with a public majority. Energy production from the publicly owned sector is 20%; whilst that from the public-private companies is 64%. The remaining 16% of total energy production is covered by about 100 private companies.

The public electricity supply utilities may be divided into:

- Generation, transmission and distribution companies
- Distribution only companies.

The first group primarily consists of the eight large interconnected utilities covering the total area of the former Federal Republic of Germany and Berlin (West). These eight utilities are the members of the Deutsche Verbundgesellschaft (DVG).

Table 8.3-1: List of the DVG utilities in the former Federal Republic of Germany

Name of Company	Annual peak load in MW in 1987	Produced energy in TWh in 1987
Rheinisch-Westfälisches Elektrizitätswerk (RWE)	23 240	118.56
PreussenElektra (Preag)	9 645	51.6
Bayernwerk (BAG)	5 300	32.3
Vereinigte Elektrizitätswerke Westfalen (VEW)	4 639	25.8
Energieversorgung Schwaben (EVS)	3 661	16 8
Badenwerk (BW)	2 692	16.9
Hamburgische Electrizitätswerke (HEW)	1 912	11.8
Berliner Kraft-und Licht AG (BEWAG)	1 830	8.6

The generation of electrical energy relies nearly totally on domestic or equivalently on secure primary energy sources. Domestic hard coal is burdened with high mining costs.

Between 1977 and 1990 the average increase in energy consumption was 2.6%, the main consumer groups being industry (45%), domestic use (29.5%), special contract supply (13%), commerce and trade (8.7%), agriculture (2.6%), public lights (0.8%).

Within Europe the German interconnected grid is strongly connected to neighbouring countries. The European interconnected system (UCPTE) connects the electric supplies of the following countries:

Netherlands	Austria
Belgium	Greece
Federal Republic of Germany	Italy
Luxembourg	Spain
France	Portugal
Switzerland	Yugoslavia.

The installed capacity within the UCPTE system was 350.9 GW in 1987. The power flow between the UCPTE countries depends on the seasons and varies very much between day and night.

8.3.2 Load-frequency control

As frequency is an indicator of the power balance in the interconnected network at any given moment frequency deviation together with any interchange power deviations between partners is taken as control criterion for maintaining equilibrium between load and generation.

A two-stage control operation for frequency and power has proved its usefulness in the interconnected West-European network.

- Primary control: by reaction from the turbine speed governors the unit output in the overall interconnected network is adjusted proportionally to the frequency deviation within a few seconds. A certain frequency deviation Δf remains.

- Secondary control (power-frequency control): deviations of the frequency and the agreed interchange power between partners are controlled by the partner in whose system the deviations occur by either increasing or decreasing the generation in the controlled power stations as required according to the system characteristic control.

The primary and secondary control functions operate in conjunction with generation scheduling and short-term optimisation (economic load dispatch).

The conditions under which power system control was initially introduced in the 1950s have changed considerably:

- The transition to larger generation units and increased use of variable pressure operation in conventional power stations has led to a deterioration in the control performance of these power plants.

- Due to various reasons, nuclear power stations still do not generally participate in primary control.

- The frequency dependent characteristic of the load has decreased.

- The ratio of the rotating masses to the prime mover power in the power stations has decreased due to less construction of turbo-generators.

The energy systems of the eight interconnected German utilities form a joint control block controlled from the control system of RWE (see Table 8.3-1) against the European interconnected system. Within this joint control block subordinate control systems maintain the exact power exchange between the different German utilities. Fig. 8.3-1 shows the different coordinated control blocks within the German interconnected system.

Stable system operation of the German interconnected power system requires effective primary control. In other words, the following requirements have to be fulfilled.

- All thermal power stations feeding into the power systems should participate in primary control, regardless of the type of primary energy.

- A band of regulation of ± 5% of the generator rating should be made available for this purpose.

- The droop characteristic should lie between 4% and 6% and the turbine governor should take action if frequency deviation occurs at a rate greater than 20 mHz (response sensitivity).

- If large, rapid frequency changes occur in the power system, the individual generating unit must react with step power changes within the band regulation with a rate of change of 0.3% P_N/second for conventional power stations and 1% P_N/second for nuclear power stations. Such rapid changes in output should be possible as often as required within a range of ± 1% without intervals.

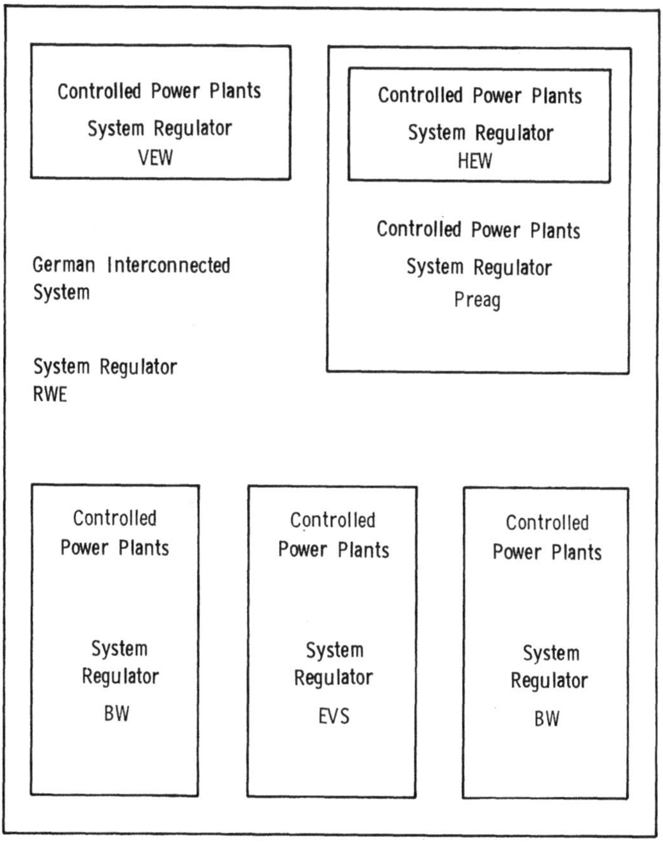

Fig. 8.3-1: Schematic representation of the power-frequency control in the German interconnected system

- For operation to a load schedule and for operation under automatic power-frequency control, generating units should be able to meet continuous power changes with rates of change of 4-12% P_N/minute. over a range of 30-60% of rated power, depending upon the type of power station, as often as required without intervals.

- Conventional power stations should be capable of delivering full output over a frequency range from 48.5 to 51.5 Hz and nuclear power stations should be able to deliver full output from 49 to 51.5 Hz. All thermal power stations should be able to deliver a least 95% of their rating at 47.5 Hz.

8.3.3 Energy management systems

Due to its specific structure each of the eight DVG companies operates its system with its own control centre. Depending on the size of the utility this system is hierarchically structured. The control of the electric power system consists of three phases:

1. Gathering information: acquisition of real-time data;
   man-machine communication
2. Decision-making
3. Action by transmitting control orders either directly (centralised remote control) or indirectly (decentralised remote control or load control).

Fig. 8.3-2 shows a schematic diagram of the functions supporting the operator in his decision-making process in the control centre. The real-time functions cover the following:

- Topological analysis
- State estimation
- Network equivalent calculation
- Security analysis
- Generation dispatch
- Voltage and reactive optimisation.

The extended real-time functions cover the following:

- Short-time load forecast
- Operator load flow
- Short-circuit calculation.

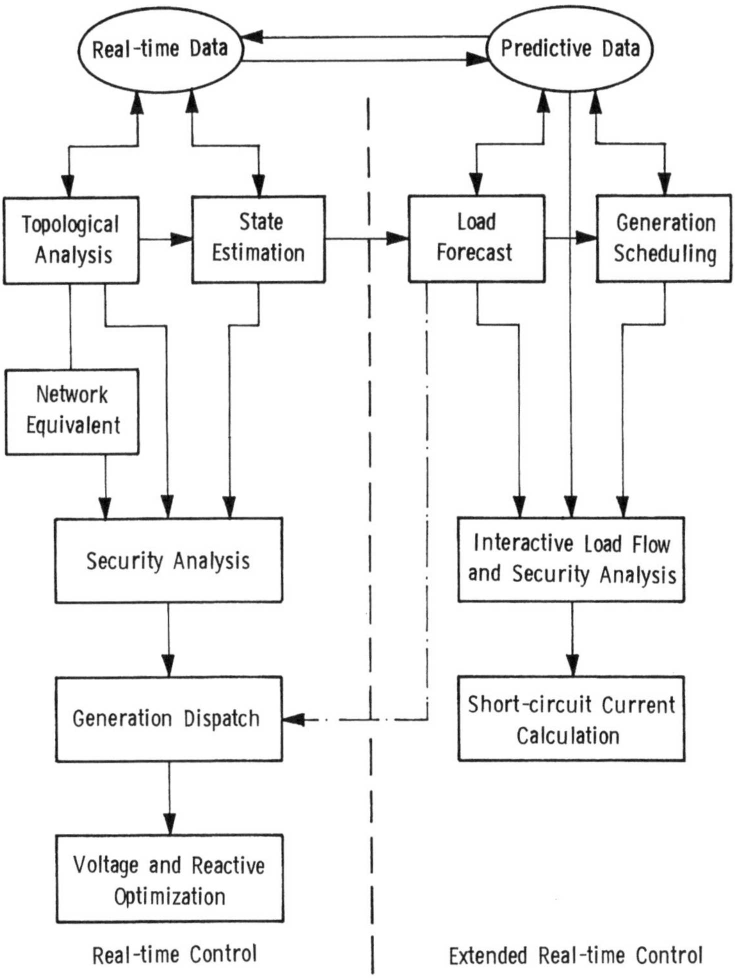

Fig. 8.3-2: Possible framework of computer aids for decision-making in control centres

8.3.4 Concluding remarks

The electric energy supply in Germany meets high technical standards. Continuous training of the operating personnel and permanent upgrading of the performance of the EMS helps maintain this standard. The following topics are of current interest for electric energy supply:

- European common market for electric energy
- Energy prices for industry
- Tariff structures
- Refund for electric energy produced by non-utility generators
- Electric heating
- Rational energy use
- Co-generation plants
- Use of alternative energy sources for electricity
- Emission control for improved environmental protection
- Use of nuclear and coal fired units for energy production.

Political changes in the Eastern European countries pose challenging new problems. The coordinated expansion of electric energy systems and information technology will contribute substantially to the urgent improvement of the electric energy supply needed in these countries.

8.4 Energy management systems in developing countries

8.4.1 Introduction

The structure of power systems in developing countries is directly related to their industrial and urban development. Large concentrations of population are common in the main cities, where development poles are located. These are reduced in number and distributed over a vast geographical area. Furthermore, generating sources are generally far away from main load centres; in many cases hydro generation is available, taking an important share of the load and energy supply.

The resulting typical longitudinal power system (LPS) is implemented through a radial configuration with generation centres electrically distant from concentrated load centres. Long transmission lines are utilised and complex operating problems encountered as a consequence. Additionally, in many developing countries, financial constraints and power system development level limit investment or delay the commissioning of new equipment. Therefore reliability is reduced and the system is in alert or a disrupted state under many operating conditions.

A LPS is very sensitive to active and reactive power changes and requires a complex coordination of active and reactive sources. Moreover, programmed and random outages may change the power system response substantially; hence close and continuous supervision is required.

The main objective of this section is to illustrate different operating problems in LPS and present the security applications required. These functions are considered basic to determine improved security control decisions. The applications presented are different from those established for highly meshed networks, and in some cases include applications commonly used in planning studies.

In general it is believed that LPS present difficult operating problems that require a good understanding of different phenomena by operating personnel. This is mainly due to the dynamic nature of the principal problems encountered during disruption.

8.4.2 Electrical characteristics of LPS

Normally an LPS can be classified as a weak system and therefore is substantially affected by nodal injections and topological changes. As the power system is continuously exposed to random variations and equipment outages, operating problems related to voltage control, frequency deviations and system instability will be common.

Short circuit capacity (SCC) can be considered as a nodal indicator of robustness. It is directly related to the equivalent positive sequence impedance seen from the node considered. In a typical LPS the equivalent impedance is dominated by the series impedance of several components, i.e. transmission lines and transformers. The SCC ranges from 500 to 2800 MVA for a typical 230 kV LPS. Typical operating problems in LPS are:

- Transmission system loadability
- Voltage control
- Transient stability
- Frequency dynamics.

A normal and continuous task in the operation of LPS is the supervision of power flows in main transmission links. However, in LPS transmission loadability is highly dependent on system configuration and the distribution of generating units.

Common transmission limitations in LPS include such factors as stability and voltage drop. Detailed load flow and stability studies are required to determine transmission limits. Power system security can be endangered when the appropriate limits are violated and power flows are not controlled accordingly. In any case, limits must incorporate the security standard for the whole system, usually determined by a single contingency. A substantial reduction in transmission loadability is determined as terminal areas are weaker. The conclusion is that in LPS transmission loadability cannot be determined based solely on a general curve and particular studies are required under different operating conditions.

Voltage control in power systems is directly related to reactive power flows and reactive power unbalances in system components. They are mainly produced by load variations and contingencies.

In LPS it is common to have load nodes electrically distant from generating sources. Hence, voltage control problems are related to the distribution and balance of reactive power in system elements.

It must be noted that any reactive power changes must be supplied locally, otherwise the voltage profile will deteriorate. Therefore operation policy should be to allocate adequate reactive power reserve margins by electric areas. Should this reserve be exhausted voltage control capability is lost and reactive power will need to be transported from distant sources, causing voltage problems. Generally coordination of reactive sources is a complex task, specially when different types of reactive power compensators are used.

Voltage control problems are more complicated to analyse when system dynamics are included. The type, speed of response, and location of reactive sources are all very important. As will be shown in the following sections, voltage collapse and stability problems are closely related in LPS.

One of the most critical problems faced in the operation of LPS is transient instability. It is known that the dynamic equilibrium of generating unit rotors depends on the ability to transmit electric power through a faulted system. In LPS characterised by weak transmitting and receiving areas, an increase in power flow causes substantial phase angle changes. Hence, any system random variation, or more important any line or generator outage, creates power unbalances that produce acceleration of the remaining units.

It is common, in LPS, to have transient instability phenomena that develop in less than a second. This obviously suggests that preventive measures must be taken to avoid these situations. Transmission limits must be assessed according to system configuration and critical contingencies considered. It must be stressed that in a daily operation topological changes are common, and different loading limits should be evaluated, through stability studies.

As mentioned before, financial constraints often limit investment and large generating plants become connected to the bulk system through transmission schemes with little redundancy. As a result, single outages can produce stability problems above certain normal generation levels. This limitation has important economical implications, from which a direct measure of security cost can be obtained. In many situations this is due to costly operating conditions, and supplementary control has to be used to increase transmission limits to maintain stability under most contingencies, and consequently making an efficient usage of available resources.

Common discrete supplementary controls used are: generating tripping, fast valving, line tripping, single phase reclosing schemes and load tripping. In all cases a security analysis is necessary to activate controls or evaluate their performance according to actual system conditions.

When the power system structure is such that a single line outage in the transmission system creates islanding conditions, detailed analysis is required to evaluate frequency behaviour and to allocate active and reactive power reserve.

As in LPS links between areas are weak, disruptions may be transmitted from area to area (with a delay) according to system configuration. As a result, dynamic behaviour in each area is different and the disruption is registered according to tie-line oscillations. Tie-line power flow changes can be obtained as a result of frequency differences between areas or phase angle displacements. As the tie-line is weak the transmitted power will be small, and larger frequency deviations are necessary to obtain a significant contribution to troubled areas. Power system response under contingencies can be analysed based on time constants of regulating controls. In analysis of LPS it is important to identify the inertial and regulating responses as they become critical in system performance.

Inertial response is obtained as a result of utilising kinetic energy from unit rotors to satisfy load requirements, consequently local frequency deviations occur. Inertial response can be evaluated assuming that turbine mechanical power changes are negligible as it depends on govenor and turbine time constants. Regulating response is obtained once mechanical power changes are developed. At this point, reserve allocation becomes important and appropriate models for turbines and govenors and limitations of generating units must be considered. As a general rule, reserve supply must be distributed according to system configuration and coordinated with load shedding schemes.

Given the sparse configuration, with LPS it is relatively common to operate in electrical islands when under disruption. Frequency rates of decay of 2-3 Hz/second can occur. This may cause frequency excursions to attain unacceptable values in less than a second. Therefore a fast and reliable load shedding scheme must be available in order to control the frequency.

8.4.3 Security assessment in LPS

From the operating problems presented in the previous section, it is clear that security analysis in LPS should include the study of dynamic phenomena, as they have great influence in system performance. In modern control centres it has become common to implement such security functions as steady state contingency analysis and automatic contingency selection. However, in LPS steady state solutions rarely define limiting conditions. These values are generally based on electrical, inertial and regulating responses. Moreover, given the structure of typical LPS, in many cases results that are relatively unimportant could be obtained, i.e. system islanding, total power flow on remaining parallel line, etc. In most cases critical contingencies can be predefined based on the system configuration.

Additionally, voltages have an important effect on active and reactive power flow under extreme loading conditions. Therefore detailed and exact calculations are required to determine voltage profile and power distribution.

Operating decisions in LPS require a set of customised computational tools, used in a preventive mode, so that security control actions can be effectively implemented. Basic security applications are presented below. As all these functions are well known, only a short review of special or important features are included.

The generalised load flow is the initial point of any dynamic study. It usually includes the following additional options:

- Voltage and frequency dependent load models
- Steady state calculation of frequency under disruption
- Generator distribution factors under outages
- Multi-island solutions
- General voltage control models, e.g. static var compensators
- Sensitivity calculations.

The definition of transmission limits and coordination of supplementary controls require a transient stability simulator with the following additional models:

- Supplementary controls models
- Automatic generation tripping
- Fast valving
- Automatic line tripping
- Automatic load tripping
- Single phase reclosing schemes
- Load shedding models
- Static var compensator models.

Frequency dynamic behaviour under disruption can be assessed using computer programs that simulate dynamic response with different model details. It is convenient to have computer programs that use a single load-generator model, with equivalent representation of electric areas and detailed generator, load and control models.

Options for the simulation of load shedding schemes, limitations of generating units and control characteristics are also very important.

The interconnection of isolated areas by a long transmission lines requires operating decisions concerning the sequence of energisation. This is specially important under very weak configurations. Additionally, switching of shunt elements may create important voltage and current transients. Main areas of study are:

- Transmission line energisation
- Shunt element energisation
- Performance of static var compensators under faults.

Most of the applications described take considerable computing resources, but most important, analysis of different cases and implementation of control actions is a time consuming process. Therefore applications are conducted off-line, but actual operating conditions, if necessary, can be obtained in data files from on-line systems.

In all cases, simulators must possess highly interactive features, specially designed for an operating environment. This is an essential factor in order to reduce computing time and to make efficient use of human resources.

An important point to stress in the design of modern control centres for LPS is to assign the same hierarchy to off-line security applications as those usually given to on-line functions, i.e. data acquisition, supervisory control, state estimation etc.

As LPS become interconnected it is most important to assess system security on a global basis, as critical problems in one system will have a substantial effect on neighbouring systems.

8.4.4 Concluding remarks

Many power systems in developing countries can be classified as LPS. Critical operating problems in LPS are related to voltage control, transient stability and frequency dynamics. Dynamic security analysis is essential in the operation of LPS. Security applications commonly implemented in modern control centres are not suitable for LPS. Due to their importance, off-line security applications for LPS must be structured with the same priority as on-line functions.

9 Project management of energy management systems

9.1 Overview

A control centre project represents a difficult task in which, very often, one or more of the following features occur:

- Project objectives are poorly defined and/or understood.
- Project deadlines are dictated by external events or arbitrarily imposed.
- Project budgets are based on naive estimates (usually underestimations).
- Project staffing is determined more by availability than ability.

As a result such a project is plagued by high development costs, delays and functional failures. With few exceptions such as control centres with simple SCADA functions, a complex SCADA/EMS system is not a standard or off-the-shelf product.

Purchase requirements should, as far as possible, match standard modules of the SCADA/EMS supplier. When new development is kept to a minimum, both customer and supplier will benefit from in-time delivery, lower costs and high quality furniture. A control centre project is an undertaking cutting across not only the boundaries of different technical disciplines, but also different groups, departments and organisations. It is often a risky joint venture involving the electric power utility, the consultant and the supplier. Hence, it is a very demanding undertaking, not only in technical but also in human terms.

The sources of control centre failure include: incorrect or incomplete system specifications; hardware failure and design errors; software design and coding errors; and human errors such as poor management. Particularly with complex control centre projects, a failure may be caused by any one combination of the above.

Since civilisation began, no other technology, apart from computer hardware, has seen six orders of magnitude in performance price gain in the space of 30 years. The cost of computer logic devices has declined at 25% per year and the cost of memory units has diminished by 40% each year.

The useful life of hardware in a control centre is less than five years, compared with about 10 years for software and 20 - 50 years for the utility's data. Often the hardware architecture and characteristics are tightly defined from the very beginning of the project, even if the functional specifications are only poorly defined.

One of the most important issues in hardware design is to anticipate:

- The future evolution of the electric power system
- The expansion of the control centre with new functions
- Technological changes and up-grade necessities.

A new control centre project implies a large amount of difficult software development. In the last decade, three factors have affected the development of software:

- The ratio of software costs to hardware costs has changed from 1:10 to 10:1.
- The cost of software maintenance is much higher than the development cost; costs appear to be about four times development costs.
- The manpower required to perform software maintenance is approximately 70% of the manpower involved in the software life cycle.

These factors explain the need for more efficient software development, requiring less maintenance.

The main reasons for the SCADA/EMS software development being so difficult are:

- The specific features of managing software
- The magnitude of the project
- Complexity/abstractness
- Software infrastructure
- Documentation
- Human interaction
- Real-time, on-line, interactive and availability constraints.

Building a large SCADA/EMS is rarely an orderly process that begins with a well defined abstract specification, continues with a smooth design phase, followed by a clean coding period and ends with a flawless implementation such as the linear model of development cycles as depicted in Figure 9.1-1.

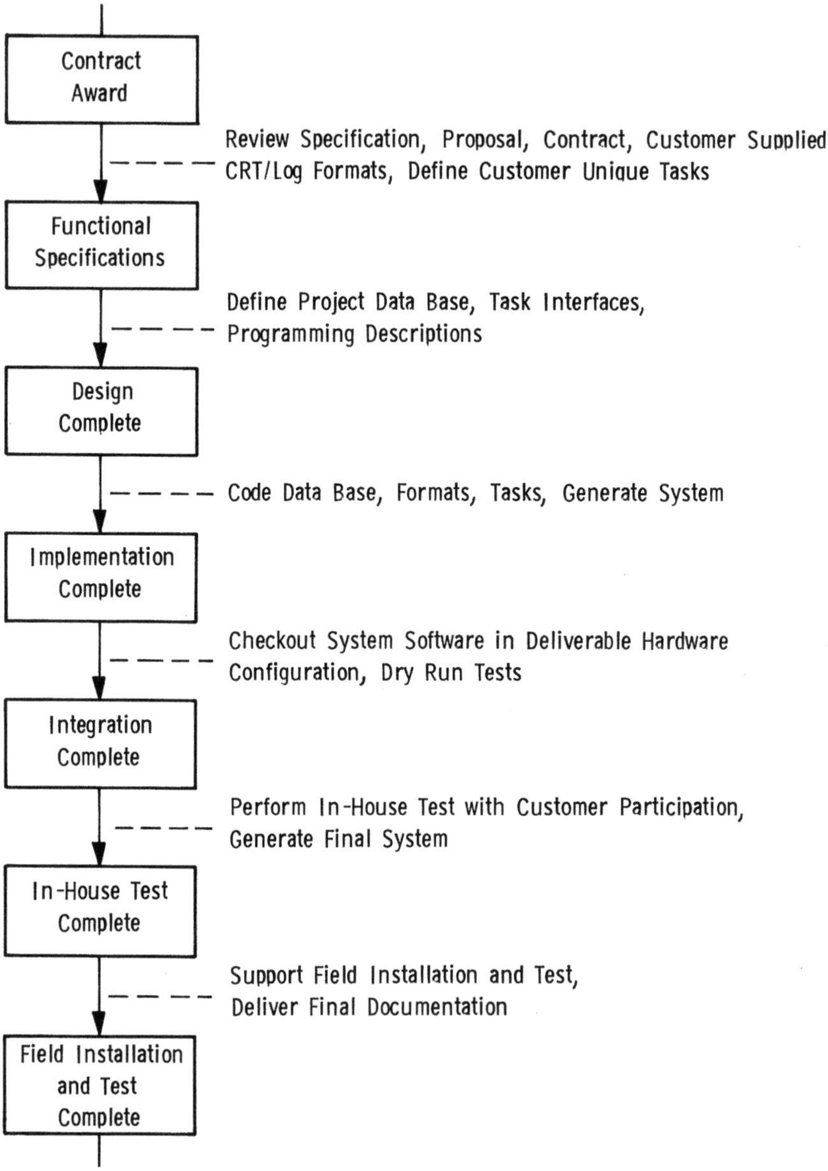

Fig. 9.1-1: A linear model of software development cycle

Software development, for complex control centre projects, is an iterative, evolutionary, spiral-like "multiple-pass" process. For a more realistic, but still simplified, model of what happens using large software, see Figure 9.1-2.

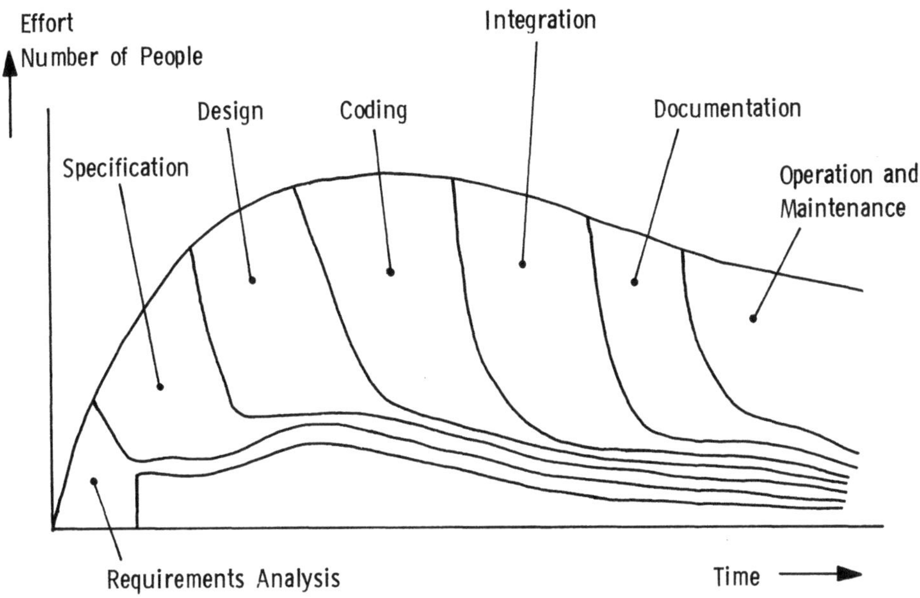

Fig. 9.1-2: A more realistic model of the software development life cycle

9.2 Stages in the implementation of a new control centre

From a customer's perspective the following steps in the implementation of a SCADA/EMS-based new control centre can be defined:

- Feasibility study
- System justification
- Functional requirements and preliminary specifications
- Releasing the specifications
- Evaluating proposals/bid evaluation
- Negotiating a contract
- Design specifications
- Implementation and organisation
- Training
- System operation.

9.2.1 Feasibility study

The first step in planning a new control system is to establish the needs, scope, objectives and benefits of the new control centre. This initial study - feasibility study - identifies the goals of the proposed system for increased security in supply and improved economic operation.

9.2.2 System justification

At this stage the utility itself determines the functions that the new control centre should perform. The result of this phase is a document that should contain:

- Functions to be performed
- Initial cost estimates
- An estimation of the time schedule (time frame of the project duration)
- Problems concerning the organisation and project management from the utility's point of view, i.e. an assessment of the organisational changes needed
- A technical/economic justification of the new control centre in order to obtain approval for the project.

9.2.3 Functional requirements and preliminary specifications

The requirements and functional specifications define:

- The tasks to be performed
- The application programs
- The support program facilities
- The computer structure model
- The data acquisition
- Communications
- Display requirements
- Interfaces with existing equipment
- Performance requirements
- System availability
- Application program execution time and priorities
- Training requirements
- Documentation requirements
- Implementation schedules
- Project management and staffing.

The functional specification is not a design specification. The document should also include such important issues as:

- Uni-supplier vs multi-supplier approach
- Functions to be developed in house
- Responsibility for system integration and maintenance
- Budgetary estimates (system cost, manpower estimates, preliminary schedule)
- In-house team.

9.2.4 Releasing the specifications

As a document the specifications should define only the portions of the system that are to be delivered by the supplier. The specifications should describe the SCADA/EMS sufficiently to ensure meaningful proposals and enable the supplier to make a realistic cost estimation. The specifications should not enter into details that would exclude the suppliers or new technical solutions. It must be stressed that this is a document at specification level and not at the design level. The specifications should remain within the functional rather than the operational level.

9.2.5 Evaluating proposals

The selection process based on the functional specification involves a detailed evaluation of the proposed system including hardware and software. The performance of the system must be measured against the objectives established. It is important also to recognise that the capabilities of the control system will have to increase to keep pace with the expansion of the power system and to accommodate new applications.

In the bid evaluation a good approach is to assess points in different areas. The proposals have to be evaluated for the full life cycle of the project from the following viewpoints:

- Investments
- Repair costs
- Costs vs benefits (for different alternatives)
- Maintenance costs
- Staff costs
- Facilities costs.

Furthermore the intangible costs of the proposals have to be carefully studied:

- Supplier experience (use of sub-contractors)
- Staff capability, longevity of staff
- References
- Maturity of the offered technology
- Obsolescence of the offered technology
- State-of-the-art and "newness" of the technology
- Supplier's financial solidity.

9.2.6 Negotiating a contract

The contract contains a statement of work prepared by the successful bidder. The statement of work defines:

- The system to be implemented
- Implementation responsibilities
- Control, coordination and project management procedures
- A description of documents to be delivered
- The project schedule.

The contract also contains the modalities for resolving exceptions and other open questions. The following commercial terms of the contract are of key importance:

- Payment deadlines
- Performance guarantees and penalties
- Warranty
- Licences
- Availability of spares
- Taxes, customs, shipping, currency
- Cancellation.

9.2.7 Design specifications

These specifications are normally developed by the supplier but power system personnel and consultants are also involved. The document describes in detail the characteristics of each of the hardware and software subsystems and related interfaces. This includes the definition of algorithms, calculation procedures, data base structures, display formats, man-machine capabilities, applications program scheduling, etc.

9.2.8 Implementation and organisation

The implementation plan and organisation should describe the tasks to be performed and the type of personnel required to perform these tasks. An effective progress reporting system should be designated.

9.2.9 Training

There are two types of training. Training for operation and maintenance is required for the effective and day-to-day use of the SCADA/EMS system. This activity allows training new staff and is required for the constant updating of the data base and the displays corresponding to the power system expansion. This kind of training is the most effective because it is on-the-job training.

Training for the future functions offers the utility staff independence and self-sufficiency in implementing customised new functions. It also permits (major) changes to the software design necessary for future SCADA/EMS upgrades.

9.2.10 System operation

System operation implies the following activities:

- An availability demonstration
- A period of continuous operation (observation and critical experience with the system)
- A warranty period.

9.3 A step-by-step plan for implementing a new control centre

From the supplier's perspective the project is divided into the following phases:

- Preliminary system design (pre-contract)
- Work statement
- System design (post-contract)
- Development
- System integration and tests
- System acceptance.

9.3.1 Preliminary system design (pre-contract)

The result of the preliminary system design phase is a document containing the following items:

- A model of the computer configuration; the exact configuration is the result of the analysis formulated with the customer
- A communication model
- Functional requirements (for each function, the response time, testing conditions, and maintenance requirements should also be included).

The document should include such issues as:

- Project management organisation
- Responsibility for system integration
- Responsibility for system maintenance
- Preliminary schedule.

9.3.2 Work statement

After the letter of intent, the project enters the contract definition phase, sometimes called the "work statement" phase. During this phase the customer's specification and the supplier's system design are unified into a single technical document - a work statement. The purpose of the work statement is to define in detail what the system does in all regards.

9.3.3 System design (post-contract)

The purpose of the system design is to define in detail how the system works. The supplier is obliged to prepare a set of documents (functional specifications) in all relevant areas for customer review. These documents define:

- The interaction of each system module with other modules
- Tables and files design
- Data base structure
- Interface specifications
- Design of input/output
- Time requirements and program timing.

A realistic time frame for the system design is between 6 and 9 months.

9.3.4 Detailed design specifications

The detailed design specification is a set of documents at the module level. At this level every module of every subsystem must be completely specified, that is:

- All internal/external data requirements are to be specified
- Testing procedures are to be defined
- Specifications are at byte and bit level of detail
- Error recovery procedures are defined.

All these documents are to be approved by the customer. This stage is the last phase before coding (for non-standard or customised functions).

9.3.5 Development

Code and unit test of software is performed during this phase. Certain project documents are now produced during this phase. These include:

- Operator's manual
- Programer's manual
- Data base modification manual
- Acceptance test procedures.

9.3.6 System integration and tests

The supplier begins with the system integration using some existing base hardware and software as a starting point. Hardware unit testing is performed as the hardware is delivered. Software unit testing is performed as each software module is completed. For the software integration the following software subsystems must be available:

- System software
- Data base structure and data entities
- Data acquisition software
- Basic display software
- Diagnostic software
- Some application software.

9.3.7 System acceptance

The system acceptance is divided into:

- Supplier site acceptance testing
- Final site acceptance testing.

The supplier site acceptance testing is performed using simulated system input/output sources. During supplier site testing the following major problems must be solved:

- System configuration and failover conditions
- Initialisation tests
- Testing under normal conditions
- Testing under abnormal (emergency conditions)
- System maintenance demonstration.

During the final acceptance testing the following major problems must be solved:

- Demonstration of functioning in an operational environment
- All the tuning activities
- Performance problems.

For factory test purposes, power system simulators are becoming available, thus making it possible to verify the full functionality of a control centre using dynamically simulated data for the entire ultimate system. This is another spin-off effect of the dispatcher training simulator.

9.4 Design, development, and maintenance of software

9.4.1 Software development phases

The following are a set of software development and maintenance guidelines intended for use by personnel involved in the power control centre industry. The phases of software development are represented in Figure 9.4-1.

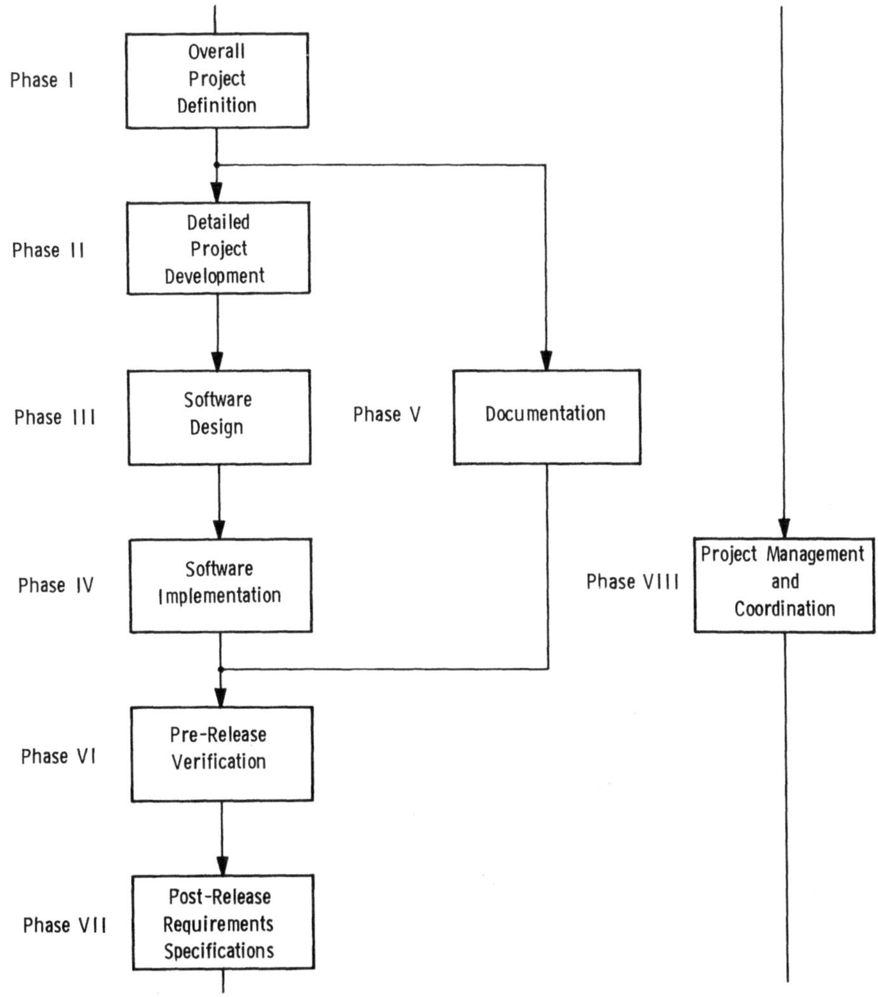

Fig. 9.4-1: The phase of software development

Phase I : Overall project definition

This phase starts with a perceived need or concept for a required capacity. The inputs include studies, requirements or need statements. The result of this phase is a report describing the overall work to be performed.

Phase II : Detailed project development

Program features and requirements are specified. The output of this phase is a detailed specification of the requirements for the program to be developed including the model specifications and any test to be used for acceptance.

Phase III : Software design

During this phase the system software is designed and test plans are specified. The results are:

- Requirements specification document
- Preliminary design document
- Detailed design document
- Test plan document.

Phase IV : Software implementation

The computer program is coded, tested and documented. The output of this phase is a working, tested code as well as the following documents:

- Program log report
- Test analysis report.

Phase V : Documentation

This phase includes the development of the various documents.

Phase VI : Pre-release verification/factory acceptance tests

The computer code, performance of the models, and program usability are verified using test cases.

Phase VII : Post-release requirements specification/site acceptance tests

Requirements for the implementation and documentation of program changes identified after program release are specified. Requirements for program maintenance are established. The output is a report describing a plan for the incorporation of modifications and improvements in the program and documentation (analogous to the requirements produced in Phase I).

Phase VIII : Project management and coordination.
This phase includes documentation, reporting, and project control including scheduling.

9.4.2 Concluding remarks

A SCADA/EMS software is a large package containing a large number of functions. The more functions to be realised, the more programs need to be written. Inevitably they will be manifold. The difficulty of software development rises non-linearly with the size and the number of programs to be written.

A SCADA/EMS software package is a large package because it contains within it a variety of technical disciplines, interests and activities. A complex SCADA/EMS lies beyond the intellectual grasp of a single individual. It will require an organisational structure and numerous people to specify, design, code, test, integrate, maintain and enhance it. The working definition currently adopted defines a software package as "large" if it requires an organisation of at least two levels of management for development and maintenance. Most SCADA/EMS suppliers have a multiple layer organisation. And finally, it is the communication between the people involved in the various SCADA/EMS of activities that leads to the size of the software development.

A SCADA/EMS package contains a large number of complex functions intricately connected and intertwined. SCADA's characteristic feature is its logical complexity. EMS's characteristic feature is its numerical complexity that requires not only "number crunching" hardware capabilities but also very refined algorithms (e.g. sparsity techniques).

The input data for the SCADA/EMS functions comes in great numbers directly from the process (on-line) and in real-time They are further characterised by inter-relationships (e.g. wrong data smearing effect) and their volatility. These features have an impact on the complexity of the software. In addition the functions are susceptible to many acceptable variants and implementation solutions. All these factors have an impact on the complexity of the software development process.

The development of customised software is a very risky undertaking for both the customer and supplier. The utility with in-house software expertise and the knowledge of its very specific functional requirements is better qualified for an in-house development. For small utilities, without in-house software expertise, such a policy is not feasible. For the supplier one major principle is: "sell standard software; don't build customised software". In developing customised software, the supplier assumes a high risk in making a price quotation and a performance assessment for an unknown software development.

The acronym "WYSIWYG" means: "What you see is what you get". Unfortunately, in software development, that is not the case. Obtaining software is not only a very complex technological undertaking but is much more difficult to understand because most of the work is "invisible". In software development, the end product is generally assumed to be a set of computer programs. However, the computer programs are only part of the delivered product. When a building is constructed, it is often necessary to build a scaffold first. When a building is finished, the scaffold is removed and all that is apparent to the observer is the building itself.

However, the construction project requires resources to erect both, i.e. the building and the scaffold. The analogy to software development is illuminating but not always obvious to senior executives. The "scaffolding" of a SCADA/EMS software development is again a software system containing a large variety of software tools. This software infrastructure, required for the SCADA/EMS software "building", is neither apparent nor "visible", although it often requires more expenditure than the product itself.

Any estimation for a control centre project which fails to estimate the resources needed (hardware, software, manpower, computer time, costs etc.) to build the software "scaffold" is not going to produce a realistic estimate. On the other hand, it is difficult for management and key executives to understand the necessity of the infrastructure and to "see" this "invisible" part of the software development.

First, it appears to cost more; the benefits of the infrastructures are not seen within the development phase. The infrastructure programs consume development resources which are scarce. It costs more in the development phase; the advantages are usually only seen by the insiders. Senior management can only vaguely appreciate what their software developers are telling them. Only very good software managers will appreciate and will invest wisely in software infrastructure. It does cost more in the development phase to build a flexible software infrastructure. However, the total life cycle is much less expensive. The fact that the SCADA/EMS software end product is not the complete measure of the work required to produce it and the necessity for the software infrastructure and accompanying overhead activities contribute to make the software-development process difficult to understand and manage.

SCADA/EMS programs are accompanied by manuals, specifications, flowcharts, listings and other documents. For every instruction, statement, line of code in the delivered package there may have been several pages of documentation containing a description of interfaces and elucidating the "why and wherefore". Donald Douglas, the airplane pioneer, stated that: "when the weight of the paper equals the weight of the plane, the plane will fly". Actually, the documentation of a Boeing weighs more than the plane. The same applies for software documentation; the only differences are that a SCADA/EMS system cannot fly and that software cannot be weighted.

However, it is difficult for some key executives to understand why for each SCADA/EMS software release thousands of pages of documentation stored on microfiches and extra megabytes memory capacity containing the software documentation are necessary. The only solution is to "educate" key executives, make them conversant with software, able to absorb and judge presentations made to them by software experts.

A software system that has a person in the loop is a much more complex system than one that does not. A SCADA/EMS system has more than one person in the loop; the power dispatch operator and the operational planning engineer must be given information in a certain way; his response must be accommodated; variability of the response must be expected.

One should also note that a SCADA/EMS system is used extensively by many users. Different users will make various demands on the computer and its software, more than an individual user. For example, a load flow program can be simultaneously used for operation, studying and training purposes from different workplaces and with different data. At the same time a network configurator could be triggered as a result of a breaker status change. All these create additional requirements in the development of SCADA/EMS software.

10 Expert systems for power system operation

10.1 Overview

SCADA/EMS provides operators with the assistance required for the economic and reliable operation of electric power systems. Within this area a large number of programs have been developed to solve important control and monitoring problems. Other problems, especially complex tasks that need human expertise, remain unsolved. The increase of interactions and the sensitivity of modern energy systems to failures tend to overburden operators. New techniques have to be adopted to coordinate the data in order that a concise picture of the current operating state of the system may be presented at any time.

Undoubtedly, a human expert will always be required in power control centres. Artificial intelligence attempts to implement expert knowledge and experience are becoming a substantial support of the power system operator in his complex decision-making process. Modern EMS concepts must aim to lessen the burden on the operator by reducing the every-day work load, but more importantly by providing decision-assisting information if necessary, and suggesting remedial measures in the case of disruptions.

Every approach to develop a decision support system integrated into the EMS has to take into consideration the following:

- A significant number of efficient algorithmic analysis programs have already been developed. The system proposed must have direct access to their capabilities.

- Artificial intelligence allows the translation of human expertise into a computer-implementable form. For computational efficiency a variety of single problem solving expert systems will need to be integrated into one coherent system.

- Communication between the operator and software tools that may help to control the power system should use a common MMI.

- The expert system needs a direct data flow to and from the on-line power system data base.

- The expert system must have real-time capability because the interpretation of incoming power system data must be made in the minimum of time to advise the operator of possible responses to system disruptions.

- The EMS must have efficient communication links between problem solving units, i.e. close coupling between different tasks of problem solving requires communication between expert systems and algorithmic solutions as well as communication between different expert systems.

10.2 Security monitoring and control

In the normal state the operator performs network security monitoring by means of on-line load flow calculations, contingency analysis etc., in order to determine preventive actions so that the power system remains in the economic and reliable operating state despite a load increase, for example. In the choice of preventive actions the operator uses his experience and intrinsic knowledge of the power system on one hand and the results of the security monitoring algorithms on the other. A similar situation exists when the operator has to choose corrective measures in order to bring the system back from the alert or even disrupted state to normally. Faced with such urgent problems he is further burdened in the decision-making process by the pressure of time. After a black-out the system restoration has to be completed as fast as possible. As each black-out has its own specific characteristics much experience is necessary to return it to a normal operating state without putting extra strength on the power system. Furthermore it should be borne in mind that large black-outs are infrequent, so that the necessary experience must be gained by training simulators.

The example of network security monitoring will be used in order to demonstrate the shortfall of today's algorithmic solutions. Network security monitoring requires analysis of present and planned operating states (after the execution of switching operations) in order to obtain complete security assessment. Short circuit programs, contingency analysis, stability programs etc. are important algorithmic tools which support the operator in the assessment and choice of the best preventive or corrective measures.

Critical evaluation of algorithmic solutions shows that even they are limited. The example of contingency simulation may be used to illustrate this point. As soon as it is possible to expand the load flow program so that it allows the computation of precisely defined contingency cases it becomes possible to get exact answers concerning the predicted operating states. The software tool best suited to solve this problem highlights the typical deficiency of algorithmic solutions. The large number of contingencies leads to a high computer load. However, improvement may be expected if the number of contingencies can be reduced to those which are really critical for the present operating state.

In general one may state that all combinatorial tasks lead to difficulties when algorithmic procedures are used for their solution. Improvement may be expected when it becomes possible to use a knowledge-based approach to select the critical cases. A further deficit of today's EMS arises when the results of algorithmic procedures have to be integrated into the decision-making process of the operator. The operator now has to evaluate the results of sophisticated analysis and prediction programs. Where technical and/or economical constraints have been violated he has to interpret the results in order to take correct decisions. As an example harmful conditioning of the model used for state estimation may be considered. Only an EMS experienced operator is able to recognise in which area of the network the state estimation results are reliable and where additional information has to be gathered in order to obtain a complete system state survey.

In order to increase the availability of the necessary knowledge and experience the heuristic procedure of the operator has to be systematised by new information processing methods. Expert systems may provide powerful tools towards the solutions for the following:

- Combinatorial problems
- Complex problems which cannot be described by one or more algorithms
- Unusual operating state where the experience of the operator is essential for the correct handling of the situation.

The introduction of expert systems into EMS is a similar innovation as the use of a data basis by means of which power industry computer applications were made possible for real-time monitoring and control.

10.3 Definitions

Research in artificial intelligence may be classified into three main areas. The first is concerned with robotics. Closely related with both robotics and the MMI in EMS is the realisation of natural language input-output-systems and picture processing. The development of expert systems (knowledge-based systems) is the third area of interest where the knowledge and experience of operators for well defined problems is used for heuristic solutions.

Expert systems are knowledge-based systems which may be used for solving problems by means of the expert knowledge stored in the computer. Expert systems assist experts by helping them to:

1. Understand the problems
2. Solve the problems
3. Explain the solution
4. Evaluate the solution
5. Enlarge the knowledge
6. Appraise the own competence
7. Structure the knowledge.

While the first two points are applicable both for algorithmic and expert system solutions the remaining items are typical for expert systems. Since only items 3, 4 and 5 are available for practical solutions the knowledge-based systems in operation today may be considered as a step towards the realisation of an expert system.

An expert system is a computer-based simulation of a human expert's solution procedure to a well-defined problem; that is the application must be structured so that development of an expert system to support the control engineer is a worthwhile task. Present day application programs (e.g. short circuit, estimation, load prediction) are characterised by short response times and limited reliability for atypical input data. The complexity of the different problems is limited. On the other hand, the work of the power system operator is characterised by its level of complexity and hence relative long response times. Problem solving depends on the experience and the knowledge of the operator. As a bridge between these two areas a specific expert system may be used to combine the algorithmic results with the operator's experience in an efficient manner.

The application of expert systems in power system operation aims to use the knowledge and experience of the operator in order to develop solutions for problems where algorithmic approaches fail or mathematical solutions cannot be obtained within the available time.

In the past the programming of well defined solutions has been prepared by formulating simple algorithms. The improvement of such programs can be achieved by more refined algorithms and the use of more powerful hardware. Today an alternative is given to the algorithmic approach by describing individual problems by means of rules, e.g. in the form of "if - then" rules. These rules have not been used in the past because solution of the problem concentrated on the development of generally applicable numerical procedures.

The solution of problems belonging to such a large and complex field as the operation of an electric power system by logical "if - then" rules is a new way of programming with the aim of optimising the available algorithmic solutions and the automation of problem solving which in the past has only been accomplished by the human operator. As the corresponding solutions are often in the form of heuristic rules, the related solution frame is called an expert system.

It is typical for expert systems that knowledge about a specific problem and the application of that knowledge for solving the problem are strictly separated. This separation offers significant advantages with respect to the development, modification and verification of expert systems. Furthermore it allows a stepwise accumulation of knowledge and experience during the use of the expert systems. Finally, it simplifies the system documentation. The new programming languages such as Prolog, Lisp, OPS5 etc. support this separation in an effective manner.

Future applications will use the principle of expert systems in order to acquire further knowledge through a knowledge base in a similar manner to that of multi-user data bases. To this end it becomes necessary to assess the competence of an expert system for the different applications. As the size of the knowledge base increases an automatic structuring of the stored knowledge becomes necessary.

10.4 Structure of the expert system

The structure of an expert system demonstrates two typical components: the knowledge basis and the inference machine. Fig. 10.4-1 shows the components of an expert system. The necessity to describe knowledge in its structure or even learning processes simply and completely by means of a rule editor highlights the difficulties encountered with the realisation of a knowledge basis in a computer. In an expert system there are three knowledge components to be administered and used:

- Domain knowledge
- Case knowledge
- Deduced knowledge

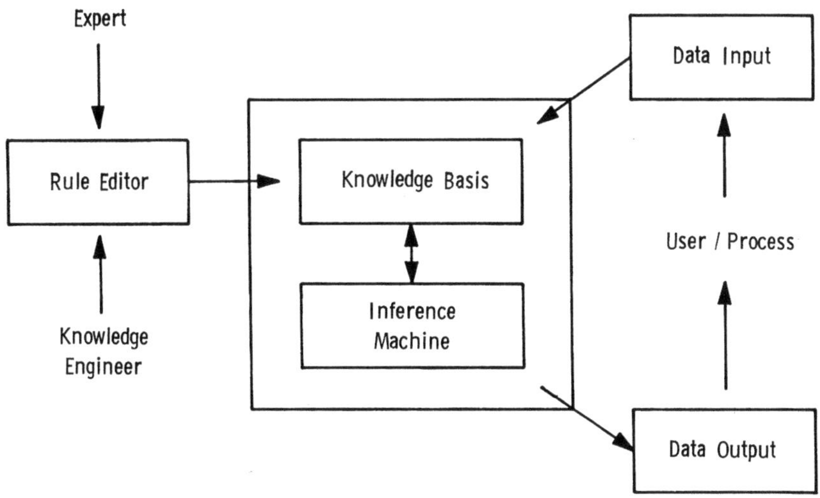

Fig. 10.4-1: Structure of an expert system

Domain knowledge comprises the knowledge of a well defined-domain. The input of the knowledge is done with the support of an expert. It has been found that the knowledge acquisition is a major problem in the construction of an expert system. The input of the rules is a time consuming task. Furthermore the formulation of heuristic knowledge requires a lot of experience on behalf of the expert providing the knowledge.

Case knowledge concerns the description of the case to be solved. The problem description by means of artificial intelligence, parameters and process variables are part of this task. During the problem description the expert system yields intermediate results which serve as deduced knowledge. It also becomes part of the knowledge basis.

Knowledge representation in the computer comprises the following three types. A Venn diagram using Boolean algebra allows the logical combination of binary information to produce a statement. Fig. 10.4-2 shows a simple example where the signals of protection relays are used for fault location. Explicit description of all relay signals belonging to a particular fault allows the automatic evaluation of alarm messages.

R7	R8	R9	...	SR7	SR8	...	Fault on Line 1 2 3, if (R7 or SR7) and (R8 or SR8)
X	X		...			...	Fault on Line 1 2 3
	X		...	X		...	
X			...		X	...	
			...	X	X	...	

Fig. 10.4-2: Knowledge representation using Boolean algebra

R7 : Relay at the beginning of line 123

R8 : Relay at the end of line 123

SRi : Secondary protection relay to R7 and R8 resp.

The explicit description of relay reactions for a given fault allows for automatic fault location.

Predicate logic is a more powerful form used for knowledge representation. Typical of this predicate logic is the separation between the verbal statements of a sentence and the relation between the different statements. As an example the sentence "Transformer T2 is the machine transformer of generator G2" takes in the predicate logic the following form: "is machine transformer of (T2, G2)".

The quantities in brackets may be variables or parameters. In the first case the predicate logic leads to a set of possible solutions; in the second it leads to a statement "true" or "false". The most important implementation of the predictive logic is the computer language Prolog.

The third form of knowledge representation is given by the production rules. Typical of these rules are the "if - then" structures. A production rule yields output data provided all the assumptions of the "if"-part are fully satisfied. Production rules are applicable until either the expected set of output data are obtained or until no further production rule is applicable. Fig. 10.4-3 shows the function of production rules where the output data a1, a2, a4, a5, a6, a7 and a9 are generated from the input data a1, a4, a5 and a7. The fourth production rule does not fire because the statement a8 is not satisfied.

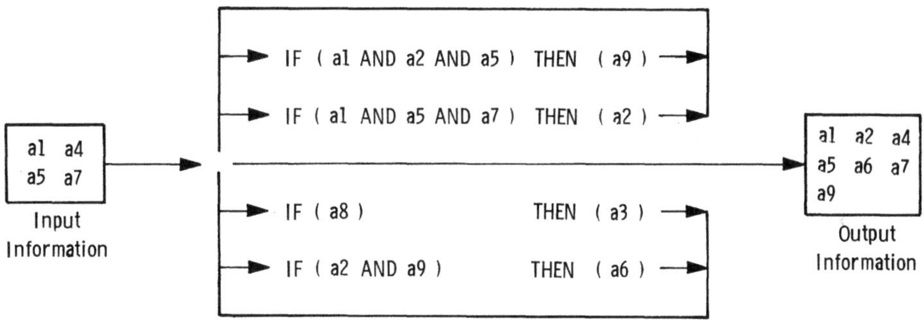

Fig. 10.4-3: Production rules for knowledge representation

After the learning phase and problem description the knowledge basis consists of a selection of rules. The use of these rules becomes possible through mechanisms which automatically combine the stored facts and rules. This part of the expert system is termed the inference machine. Depending on the chosen strategy (forward or backward chaining) the inference machine searches for the case specific rules within the knowledge basis in order to answer questions of the operator. As shown in Fig. 10.4-4, the inference machine may lead to intermediate results before the final decision can be taken by the expert system. Further components of the expert system are the rule editor and the input and output machines (see Fig. 10.4-1). The input machine supports the operator in the problem description. This process is comparable to the learning phase of the expert system.

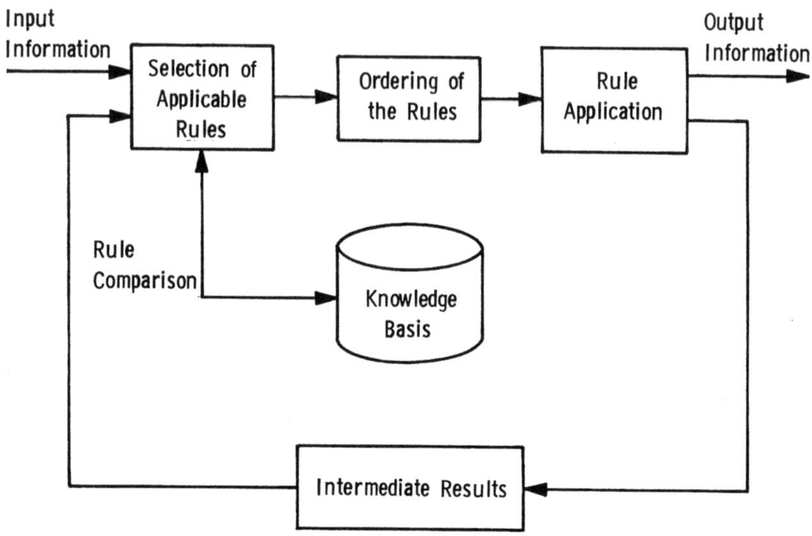

Fig. 10.4-4: Inference machine evaluating the production rules

Very important in this respect is the interface to the on-line data base of the SCADA/EMS in order to have fast and direct access to all the data describing the operating state of the power system.

Each decision may be confirmed by the expert system on request. To this end the used rules are summarised and displayed in a suitable form. Questions concerning "why" or "how" may be answered in this manner.

The usefulness of any technical system depends very much on the availability of the stored knowledge. Interfaces to the on-line data base and to process communication links are necessary for expert systems to be used in EMS.

A data basis is a good example to show how factual knowledge is implemented. The mechanism to use the data base knowledge is a simple procedural knowledge representation. Artificial intelligence is characterised by the fact that solutions to problem are not only obtained by analysing factual knowledge. The reason for this is the complexity of problems, i.e. an appropriate set of rules could be far too large. A second reason is the fact that the data base would have to contain elements which are not known when the solution to the problem is defined. It is the derivation of new facts (deduced knowledge) using available information and present data which principally characterise the use of knowledge- based systems.

The computer based formulation for any heuristic knowledge available when the solution is specified plays an important part in the construction of an intelligent problem solving mechanism (knowledge engineering). In this respect the following two questions have to be answered:

- What is the best knowledge representation for a given problem?
- What is the optimal inference mechanism to process the information in the knowledge basis?

The answer to these questions very much determines the applicability and usefulness of an expert system for EMS.

10.5 Possibilities and limits of expert systems

Three important properties of an expert system will be illustrated by practical application:

- Object oriented programming
- Representation of heuristic information
- Interpretation of results.

Important amongst the new programming languages is (Prolog) based on a concept which itself is a relational data base. Hence any type of data structures may be defined by the user. The access to stored information is simple and comfortable. The information basis can be realised so that assumptions and constraints of a specific case are only requested when required. Advantages of the data administration within declarative programming allow for a clearly arranged representation of the electrical components of the power system together with their static and dynamic parameters and associated measurements. The automatic limit checking of individual data can easily be realised.

The development of algorithmic programs requires a complete problem description, e.g. by algebraic equations. A large number of "if - then" statements is a disadvantage in such a program. On the other hand, artificial intelligence uses incomplete knowledge for problem-oriented solutions.

An important property of the formulation of computer based solutions using partial solutions is the fact that the final decision may be justified through rules used during the inference process. Explanation of the results of an expert system is a direct consequence of the explicit description of knowledge gained from the solution .

Expert systems are still limited by the following problems:

- Limited possibilities for numerical calculation within the expert system
- The interface to algorithmic programs is time consuming
- The implementation of the expert system has a large influence on the execution time
- Storage requirements are large
- The complete, accurate and consistent knowledge basis requires a lot of development.

10.6 Applications

Modern power systems are operated by highly skilled operators supported by computerised control systems. The EMS is the centre of a network control system organised through various levels of computer processing. The object of the EMS is to ensure the secure and economic operation of power systems. The traditional numerical approach is mainly designed for dealing with the normal state, being less effective during emergencies. In such situations, the solution rests heavily upon the experience and skills of the operator. Coping with emergency events requires diagnosis and decision processes supported by more sophisticated software. Typical applications are:

- Alarm processing and fault diagnosis
- Stationary and dynamic security analysis
- Determination of predictive or corrective measures
- Voltage/var control
- Pattern recognition in security assessment
- Power system restoration.

In view of these applications the distinction between algorithmic and logical problems is helpful:

- Algorithmic problems are suitable for numerically defined tasks which have an iterative or recursive solution, e.g. optimisation procedures
- Logical problems are suitable for combinatorial or topological tasks which have a non-numerical solution.

Based on the concept that expert systems will be useful for all kinds of problems where some interpretation, combination or evaluation of results is necessary many new fields of application within an EMS can be foreseen in the realisation of MMI.

Fig. 10.6-1 shows a simplified structure of present day EMS. Many powerful algorithmic programs (power application software) are available to the operator in the form of state estimation, on-line load flow, short circuit analysis, security assessment.

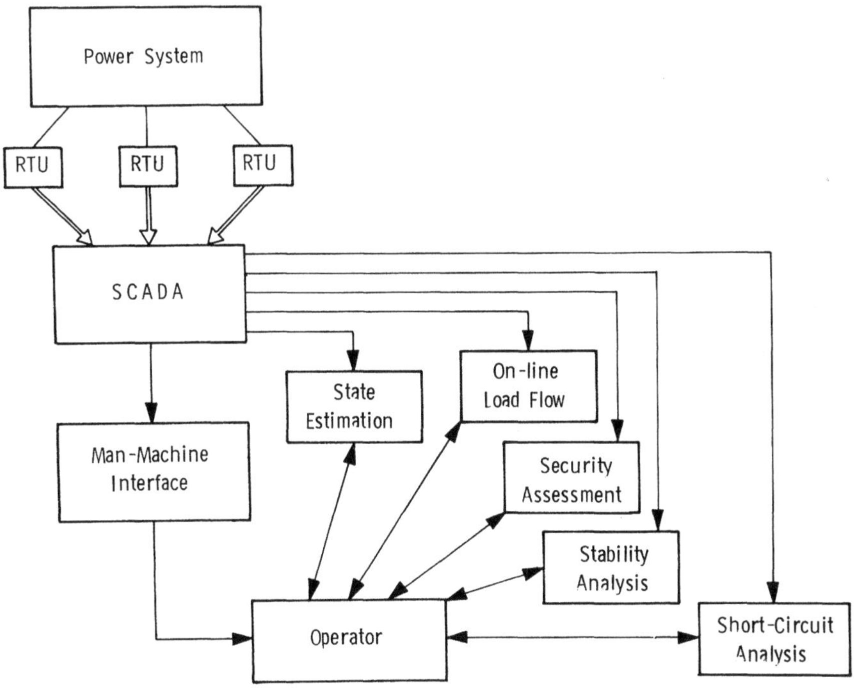

Fig. 10.6-1: Power system control using power application software

For effective power system monitoring and control the operator has to decide which of these programs he uses in any given situation. Especially in heavily disturbed situations it is difficult to decide which program to use in order to find a problem-oriented solution as fast as possible.

The introduction of expert systems in EMS allows for the combination of algorithmic and heuristic procedures. The storage of skill and experience into the integrated solution leads to a new EMS, as shown in Fig. 10.6-2. Here the expert systems support the operator in the choice of the optimal EMS tool in each specific state of the power system. The operator no longer has to decide which algorithmic tool to use in order to find the correct operational decision. Hence, the intelligent MMI realised as an expert system assists the operator in selecting which software tool of the EMS to use.

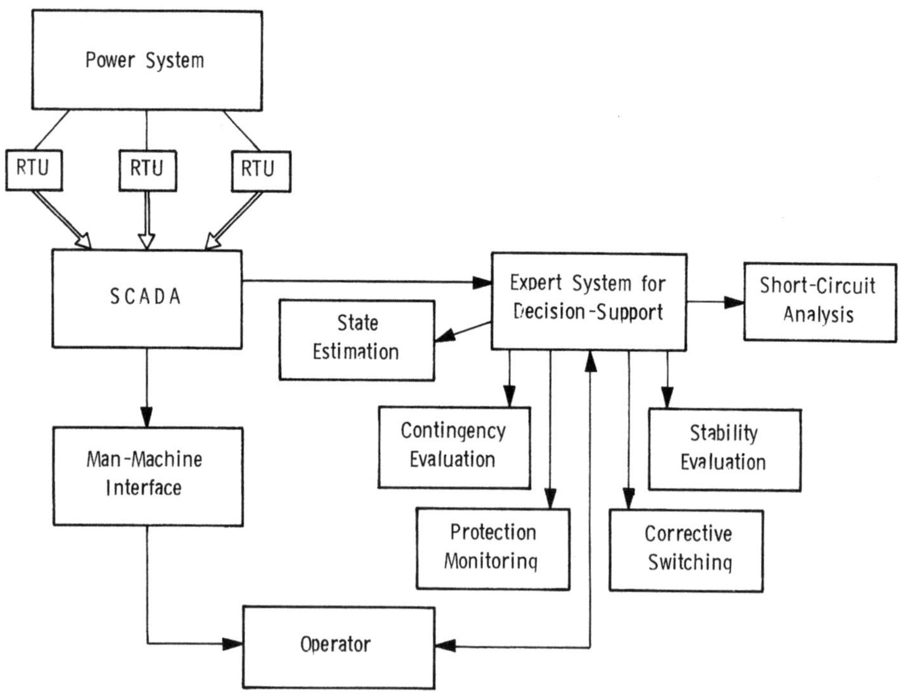

Fig. 10.6-2: Power system control using an EMS with an integrated expert system

10.7 Conclusions

Expert systems will provide interesting and new solutions for modern EMS machines of the future. For the development of new applications the following three steps have to be performed:

- Construction of a prototype
- Integration of the prototype into the on-line EMS
- Maintenance and expansion of the expert system.

In the area of power engineering of expert systems there are still many research and development problems to be solved. Basically there are three classes of problems:

- Maintenance and expansion of large knowledge basis
- Real-time properties of expert systems
- Automatic knowledge acquisition.

In solving these problems the classical separation into research and later application is no longer feasible. A close cooperation between the developer and the user of expert systems is mandatory. Subproblems such as knowledge representation or data interfaces must be coordinated and solved using the knowledge and experience of all the people involved in order to obtain practical solutions.

The topic of expert systems has grown rapidly during the last few years. It offers new and interesting possibilities in power engineering. For power system applications the available results have to be carefully analysed and developed further taking the specific constraints of EMS into consideration. This is the only way to achieve solutions which are useful for practical applications. A good understanding of expert systems and the intrinsic knowledge of power systems are important prerequisites for the successful transfer of this new information technology into power system operation.

For the evaluation of expert systems the following aspects may be useful. Artificial intelligence has initiated a new approach for program development due to the creation of new languages and knowledge representation. Through the use of predicative, knowledge-based or objective-oriented languages equivalent alternatives are available to the algorithmic solutions used so far in EMS.

Expert systems are not to be seen as a possible replacement of algorithmic solutions but as a supplement and extension. They will be used for the solution of those problems for which so far no complete or satisfactory solution based on an algorithm has been found. The use of expert systems for power system monitoring and control will greatly effect future EMS towards increased flexibility and ability. When expert systems are initially installed the best procedure is to use additional hardware interfaced to the on-line EMS.

The close collaboration between utilities, suppliers of EMS and research institutes is the key to a successful development of new EMS using expert systems.

Glossary

The SCADA/EMS field has rapidly expanded over the past several years; as a result, new terminology is appearing at a phenomenal rate. However, network control technology is still a new field.

The literature reflects its immaturity in two ways. First the annoying tendency for specialists to use entirely different terms for the same concept, or, perhaps worse, for specialists to use the same word to mean entirely different things.

This glossary is compiled from information acquired from numerous books, journals and authorities in the field of power dispatch and computer technique.

The purposes of the glossary are:

- To document current use of the corresponding terminology
- To encourage standardisation among competing and conflicting terms.

The authors hope that this compilation will help clarify the terminology in the field of network control; nevertheless because another definition for a glossary is: "a list of mystifying definitions not containing the word you seek" the authors apologise for any omission.

ACCESS TIME: The time interval between the instant at which data are called for from a storage device and the instant delivery is completed.

ADAPTIVE CONTROL: A control method in which automatic means based upon the monitoring of independent variables are used to change the control parameters in a way intended to improve the performance of the control system.

ADDRESS: A number specifying where a unit of information is stored in the computer's memory.

ALGORITHM: A step-by-step procedure that has a specific beginning and end and is guaranteed to solve a specific problem. A completely specified procedure for performing a computation in a finite amount of time.

ANALOG COMMUNICATION: Transfer of information a continuously variable quantity, such as voltage produced by a strain gauge or air pressure in a pneumatic line.

ANALOG DATA: Data represented by a physical quantity that is considered to be continuously variable and whose magnitude is made directly proportional to the data or to a suitable function of data.

ANALOG INPUT: An electrical input of variable value provided to the energy management system by a sensing device.

ANALOG TO DIGITAL CONVERTER: A circuit or device whose input is information in analog form and whose output is the same information in digital form.

APPLICATION SOFTWARE: Software specifically produced for the functional use of a computer system; for example, power application software, in contrast to system software.

ARIMA-Model: Autoregressive, Integrated, Moving-Average. A model for a nonstationary, nonseasonal series containing both regular auto-regressive and regular moving-average parameters, where the series has been differenced to achieve stationarity.

ARMA-Model: Autoregressive, Moving-Average. A model for a stationary, nonseasonal series containing both regular auto-regressive and regular moving-average parameters.

ARTIFICIAL INTELLIGENCE: The subfield of computer science concerned with developing intelligent computer programs which includes programs that can solve problems, learn from experience, understand language, interpret visual scenes, and, in general, behave in a way that would be considered intelligent if observed in a human.

ATTRIBUTE: A simple property attached to an object. In production systems the data memory may be represented as a set of attribute-value elements.

AUTOMATIC ENERGY CONTROL: Inadvertent energy exchange pay back.

AUTOMATIC GENERATION CONTROL/POWER SYSTEM FREQUENCY AND TIE-LINE CONTROL: Control of frequency and tie-line power in a power system. The regulation of the power output of electric generators within a prescribed area in response to changes in system frequency, tie-line loading, or the relation of these to each other, so as to maintain the scheduled system frequency and/or the established interchange with other areas within predetermined limits. Control algorithms which are executed cyclically in the real-time as part of a closed loop control without interaction of the dispatcher.

AUTOMATIC LOAD CONTROL: Control of loads, e.g. by ripple control.

AUTOMATIC VOLTAGE CONTROL: Control of voltage and reactive power production.

AVAILABILITY: The probability that an item is functioning satisfactorily at any instant of time when used according to designed conditions.

BACKTRACKING: A control structure of AI-computer programs that is characterised by returning the database, or conditions, in a process of problem solving to a previous state in order to try an alternative solution path (see control structure).

BACK-UP: Provisions made for the recovery of data or software, for restart of processing, or for the use of alternative computer equipment after a system failure.

BACKWARD CHAINING: A problem-solving principle that attempts to solve a problem by stating a goal and looking in the knowledge base for the conditions that would cause it to come about. Afterwards reiterating this process, using those conditions as the goals and searching for their preconditions, and so on (see forward chaining).

WRONG DATA: Term applied to measurements that deviate from the time value by at least 5 times the variance associated with that measurement.

BATCH MODE: A mode of execution of user unitiated tasks without intermediate user intervention.

BATCH PROCESSING: The practice of running a stream of programs through the computer so that each is executed from start to finish without interruption, i.e. as opposed to multiprogramming or time sharing.

BAUD: A unit of speed for data communications, equal to the number of times per second a signal is altered. Although baud is usually equivalent to bit rate, coding techniques make it possible to send more than 1 bit at a time in some systems.

BLACK START: Reconstruction of power system after major disruption.

BLACKBOARD: A system architecture that employs a data base or memory that is accessible to several processes, called knowledge sources. The memory that is common to all processes serves as a basis for communication of intermediate results among rules or knowledge sources.

BLACKOUT: Complete energy supply interruption within a certain area or an entire utility.

BOTTOM-UP: Pertaining to an approach that starts with the lowest level software components of a hierarchy and proceeds through progressively higher levels to the top level component; for example, bottom-up design, bottom-up programming, bottom-up testing. Contrast with top-down. A strategy of proceeding from the simple and concrete to the complex and abstract.

BREADTH-FIRST SEARCH: An approach in which, starting with the root node, the nodes in the search tree are generated and examined level by level before moving to the next level.

BROWNOUT: Partial energy supply interruption within a certain area or an entire utility.

BUFFER: A storage device used to compensate for difference in rate of flow of data or time of occurrence of events when transmitting data from one device to another.

CENTRAL CONTROL: A system whereby all data analysis and control actions are performed at a single point.

CENTRAL PROCESSING UNIT (CPU): The part of the computer that performs the interpretation and execution of instructions. It does not include memory or input/output functions.

CHI-SQUARE STATISTIC: In estimation for Box-Jenkins models the chi-square statistic is used to test whether a set of residual auto-correlations up to some lag k reveal any evidence that the residuals are significantly correlated.

CLOSED-LOOP: A system in which the value of some output quantity is controlled by feeding back the value of the controlled quantity, and using it to manipulate an input quantity so as to bring the value of the controlled quantity closer to a desired value. A control system in which output data are measured and fed back to the control for comparison with the input data, so that information flows back and forth.

COMPLEXITY (SOFTWARE): The degree of complication of a system or system component, determined by such factors as the number and the intricacy of interfaces, the number and intricacy of conditional branches, the degree of nesting, the types of data structures, and other system characteristics.

COMPUTER NETWORK: A complex consisting of two or more interconnected computers.

COMPUTER-BASED SYSTEM: An energy management system in which a computer is the central controlling device.

CONFIDENCE LEVEL: Statistical conclusions are not generally made with absolute certainty. When a statistical hypothesis is concluded to be true on the basis of some statistical test, the conclusion is usually stated in terms of a degree, or level, or confidence expressed as some percentage.

CONTINGENCY ANALYSIS: The prediction of systems conditions as a result of some outage event. The goal of the CA function is to plan a defensive strategy to counter the expected difficulties. Calculation of the effects of an outage of an arbitrary element within the power system.

CONTROL STRUCTURE: A construct that determines the flow of control through a computer program. A problem-solving strategy determined either by the machine (von Neumann architecture) or by the program. The control structure means the strategy for manipulating the domain knowledge and the data in order to solve a problem.

CONVERSATIONAL: Pertaining to an interactive system that provides for interaction between an user and a system similar to a human dialog.

DATA: A representation of facts and concepts in a formalised manner suitable for communication, interpretation, or processing by human or automatic means; raw data gained from the power system elements such as status of a breaker, measured analog values, etc. (Data vs. Information).

DATA ACQUISITION SYSTEM: A centralised system which receives data from one or more remote points. Data may be transported by either analog or digital telemetering.

DATA BASE: A set of data, part or the whole of another set of data. A collection of data about objects and events on which the knowledge base will work to achieve desired results. A relational data base is one in which the relationships between various objects and events are stored explicitly for flexibility of storage and retrieval.

DATA BASE MANAGEMENT SYSTEM: A program that enables a data base to be organised to expedite the sorting, updating, extracting, or retrieving of information and the generation of reports or desired output.

DATA ITEM: The smallest externally visible unit of data in terms of which the records, areas/files of a collection of data are constructed.

DATA STRUCTURE: A formalised representation of the ordering and accessibility relationships among data items without regard to their actual storage configuration.

DEBUGGING: A procedure that executes whenever a particular predicate about a database becomes true.

DECLARATIVE KNOWLEDGE REPRESENTATION: Knowledge about the solution of a certain problem has to be coded in order to prepare the computerised problem solving process. AI-techniques provide different new concepts for representing such knowledge that contrast to the well known algorithmically way of coding knowledge used to implement number crunching programs like load flow or estimation. Probably the most important concept is the declarative knowledge representation that describes domain knowledge in the form of facts and assertions (see knowledge representation).

DEPTH-FIRST SEARCH: In a hierarchy of rules or objects, depth-first search refers to a strategy in which one rule or object on the highest level is examined and then the rules or objects immediately below are examined.

DIGITAL INPUT: An electrical input of one or more bits provided to the energy management system.

DIGITAL TRANSMISSION: A method of sending and receiving information as a code made up of on-and-off pulses of electricity or light.

DIGITAL-TO-ANALOG CONVERTER: Interface to convert digital data into analog data. Device for translating a digital quantity, represented as a number of binary units, into a voltage or current whose magnitude is in some way equivalent.

DISPATCHER (SYSTEM OPERATOR): The person or persons charged with operating the EMS, and therefore handling the power system controls, planned outages, and potential transactions with other utilities.

DISPATCHING: The overall real-time control of the power system involving such operations as: the assignment of load to specific generating stations and other sources of supply to effect the most reliable and economical supply as the total load rise or fall; the control of operations and maintenance of high-voltage lines, substations, and equipment, including safety procedures; the operation of principal tie-lines and switching; the scheduling of load reduction through the use of a load management system and load shedding devices.

DISTRIBUTED CONTROL SYSTEM: A network of independently functioning controllers under the jurisdiction of a central control element having the capability at each controller of data analysis and control action.

DISTRIBUTION AUTOMATION: A means of performing various functions on one or more distribution lines, within the confines of the low voltage side of the distribution substation via a communication link. Distribution automation includes all communication and control functions for the distribution system, such as: load control; remote meter reading; feeder switching; capacitor bank control; transformer (temperature) monitoring; fault location and isolation.

DYNAMIC NETWORK COLOURING: Dynamic colour displays are provided to effectively show the electrical state of the network. The function "informs" the operator which line or network segments are isolated or energised.

ECONOMIC DISPATCH CALCULATION: Optimal distribution of the load on the generators in operation, calculation of set points and participation factors for automatic generation control.

ECONOMY ANALYSIS: Functions for the optimal cost effective operation of a power system.

EFFECTIVENESS: The capability of the item to carry out its assigned mission.

ENERGY MANAGEMENT SYSTEM (EMS): Computerised control system for real-time control, security analysis and economic optimisation of a state-wide power generation and transmission network. A system, the expressed purpose of which is to control the consumption of, the demand for, or the supply of energy. A term that denotes that part of electric power utility system operation that plans, coordinates, and controls power supply, transmission, distribution, and utilisation.

ENTITY: An object, person, placer or event in the real world which relates to data in a database.

EQUIVALENT: Calculation of network equivalents on the active and passive influence of unmeasured neighbouring power systems.

EQUIVALENT NETWORK: A network which may replace a given network without changing the state variables at specified boundary nodes.

ERROR DETECTING/CORRECTING: The use of error-detecting codes so that a character or block detected as being in error automatically initiates a request for retransmission of the signal detected as being in error or reconstructs the original character or block from an error correction algorithm.

ETHERNET: Local network for sending messages between computers by way of a single coaxial cable.

EVOLUTIONARY DEVELOPMENT: The practice of iteratively designing, implementing, evaluating, and refining computer applications.

EXPERT SYSTEM (XPS): A computer program that uses the representation of domain knowledge and a broad variety of reasoning techniques to solve problems that normally require the abilities of human experts. An expert system typically consists of two strictly divided components. The knowledge base loaded from an editable knowledge base file containing domain and application specific knowledge, e.g. coded in expression of first-order predicate logic. The second component of an expert system is the inference engine that applies the knowledge provided by the knowledge base in order to find solutions to user defined problems. The inference engine contains no application specific information, therefore it might be used in quite different application domains determined by the actually loaded knowledge base file (see knowledge base and inference engine).

EXPERTISE: Proficiency in a specialised domain. An expert system is said to have expertise in its domain if its performance is comparable to that of a human with 5 to 10 years of training and experience in the domain. Expertise often consists of massive amounts of information combined with rules of thumb, simplifications, facts, and wise procedures in such a way that one can analyse specific types of problems in an efficient manner.

EXPLANATION FACILITY: Because of the explicit and the most declarative representation of domain knowledge within an expert system it is possible to provide reasons for a solution found by the inference machine. The explanation facility describes that part of an expert system that tells what steps and what knowledge were involved in the processing by which it arrived at a solution.

EXTENDED REAL-TIME FUNCTIONS: Predictive or prophylactic functions/ advanced power application functions. Computer aided before the fact decisions by the operating personnel.

FAILURE: The termination of the ability of a unit to carry out its assigned mission.

FAULT: An attribute which adversely affects an item's reliability.

FAULT IDENTIFICATION: Function of the distribution automatisation used to locate primary faults.

FAULT-TOLERANT COMPUTING: The ability to execute given algorithms successfully regardless of computer software errors and hardware failures.

FIBRE OPTICS: A communication technique where information is transmitted in the form of light over a transparent fibre material such as a strand of glass. Advantages are noise-free communication not susceptible to electromagnetic interference.

FIFTH-GENERATION COMPUTERS: The next generation of computing machines. It is assumed that they will be larger and faster and will incorporate fundamentally new designs. Parallel processing, the ability of a computer to process several different programs simultaneously, is expected to result in a massive increment in computational power.

FILE: A collection of data records treated as a single unit.

FIRST-ORDER PREDICATE LOGIC: First-order predicate logic permits assertions to be made about variables in a proposition. Therefore it is a popular form of logic used by the AI community for representing knowledge and performing logical inference (see knowledge representation).

FORECAST: A numerical prediction for a time series of what happen at some future period, along with an assessment or measurement of the uncertainty in the prediction.

FORECASTING OR PREDICTIVE DEMAND CONTROL: A demand control algorithm where the demand is calculated for the demand interval based upon the history, incremental rate of usage, and the time remaining in the interval.

FORWARD CHAINING: Problem-solving principle characterised by working forward from known facts toward conclusions. The method that starts with initial knowledge and applies inference rules to generate new knowledge until either one of the inferences satisfies a goal or no further inferences can be made. In forward-chaining production systems, the applicability of a rule is determined by matching the conditions specified on its left-hand side against the knowledge currently stored in data memory. An expert system that employs forward-chaining techniques is also called a data-driven system.

FRAME: A knowledge representation based on the idea of a frame of reference. A frame carries with it a set of slots that can represent objects that are normally associated with the subject of the frame. The slots can then point to other slots or frames.

FULL DUPLEX: A communication channel that transmits data signals in two directions simultaneously.

FUNCTIONAL DECOMPOSITION: A method of designing (conceiving) a system by breaking it down into its components in such a way that the components correspond directly to system functions and subfunctions.

FUNCTIONAL REQUIREMENT: A requirement that specifies a function that a system or system component must be capable of performing.

FUZZY LOGIC: An approach to approximate reasoning in which truth-values and quantifiers are defined as possibility distributions that carry linguistic labels, such as true, very true, not very true, many, not very many, few, and several. The rules of inference are approximate, rather than exact, in order to better manipulate information that is incomplete, imprecise, or unreliable.

GARBAGE COLLECTION: A technique for recycling computer memory cells no longer in use.

GAUSSIAN: A Gaussian distribution is a frequency distribution for a set of variable data, sometimes called a normal distribution and typically represented by a bell-shaped curve that is symmetrical about the mean.

GENERATION SCHEDULING AND CONTROL: A set of EMS functions that regulates the real power output of generators, calculates the desired real power output for on-line generators, and provides dispatcher file management for scheduling power interchange to and from neighbouring utilities. The scheduling of generation facilities for a specified period.

GOAL DRIVEN: A problem-solving approach that works backward from the goal. A top-down approach often referred to as "hypothesise and test".

HALF DUPLEX: A communications channel that transmits data signals in two directions, but not simultaneously.

HEURISTIC: Experimental, judgmental knowledge; the knowledge underlying "expertise"; rules of thumb, rules of good guessing, that usually achieve desired results but do not guarantee them. A process, sometimes a rule of thumb, that may help in the solution of a problem but does not guarantee the best solution, or indeed, any solution. The primary effect of heuristics is to eliminate the need to examine every possible approach. Knowledge derived from experience. One of the major obstacles in providing machines with intelligence is the current inability to define and structure heuristics for computer programming.

HIERARCHICAL DECOMPOSITION: A method of designing a system by breaking it down into its components through a series of top-down refinements.

HIGH-LEVEL LANGUAGE: A language in which the instructions more closely resemble English. One high-level language instruction is often converted into several machine-language instructions.

ICON: An image-like symbol to which a computer user can select an interface device in order to choose a function, such as "move window".

INFER: To derive by reasoning; to conclude or judge from premises or evidence.

INFERENCE: The process of reaching a conclusion based on an initial set of propositions, the truths of which are known or assumed. A process by which new facts are derived from known facts.

INFERENCE CHAIN: The sequence of steps or rule applications used by a rule-based system to reach a conclusion.

INFERENCE ENGINE: The most important part of an expert system besides the domain specific knowledge base. The application independent inference engine uses the knowledge represented in the knowledge base. In order to find the solution to problems the inference engine selects and infers rules together with the data describing the actual problem (see expert system).

INFORMATION: Data which has been transformed, converted, related, compared, computed. (Information vs. Data)

INTERCHANGE TRANSACTION EVALUATION: Determination of the economies of interchange of energy between utilities.

KNOWLEDGE: The information a computer program must have to behave intelligently. Facts, beliefs and heuristic rules. An integrated collection of facts and relationships that, when exercised, produces competent performance. The quantity and quality of knowledge possessed by a person or a computer can be judged by the variety of situations in which the person or program can obtain successful results.

KNOWLEDGE ACQUISITION: The process of extracting domain knowledge from domain experts. The process of incorporating domain knowledge into an expert system by extracting it from domain experts and encoding the information into an internal representation, such as rules.

KNOWLEDGE BASE: The part of a knowledge-based system that contains domain specific knowledge and heuristics that should be applied by the inference engine. The data base is made up not merely of files of uniform content, but of facts, inferences, and procedures corresponding to the types of information needed for problem solution. The knowledge base of expert systems is typically exchangeable via files in order to feed the inference engine with knowledge of different applications (see expert system).

KNOWLEDGE ENGINEERING: The work of software engineers who acquire knowledge for knowledge-based systems and decide how to represent it for use in the system. Two important aspects of the task of knowledge engineering should be mentioned. On one hand it deals with questions of knowledge acquisition, e.g. interview techniques in order to extract the expertise in order to describe the relevant knowledge. On the other hand, the knowledge engineer has to decide how to represent the expert knowledge and how to use it in order to achieve an optimal performance of the expert system (see knowledge representation).

KNOWLEDGE REPRESENTATION: In order to automate the application of expert knowledge it must be transformed into a more structured form capable of being understood by a computer. Knowledge representation means a structure in which knowledge can be stored in such a way that allows the user and the system to understand the relationship between areas of knowledge and to manipulate those relationships (see first order predicate logic).

LAYERING: A method of logically organised data in a data base. Functionally different classes of data are segregated on separate layers, each of which can be displayed individually, or in any desired combination.

LIMIT CHECK: Determination of those components with operating conditions exceeding the limits.

LISP: A programming language designed specifically to manipulate symbols rather than numeric data.

LOAD BUS: A node with predetermined active and reactive power input (PQ-Bus).

LOAD FORECAST: An estimate of the expected load of a network at a given future date.

LOAD FORECASTING: Load forecast, based on measured load curves considering climatic effects (temperature, humidity, etc.).

LOAD MANAGEMENT (User): As a part of indirect load management, the automatic monitoring and control by a user of his loads to take advantage of the electric rate of economic incentives. Load management is the action(s) initiated by the utility or one or more of its customers as a result of financial incentives to: control load growth; alter the shape of the load curve; increase supply through non-utility or non-traditional sources. The action(s) can be initiated to: reduce capital expenditures; improve capacity limitations; provide for economic dispatch; reduce the cost of service; improve load factor; improve system efficiency; improve system reliability.

LOAD SHEDDING: The process, implemented by load controllers or demand controllers, of user removal of preselected loads.

LOAD STABILITY: The capability of a load consisting of rotating machines to regain a steady state of operation after a disruption.

MASTER: The entire complement of devices, functional modules, and assemblies which are electrically interconnected to effect the master station supervisory functions. The equipment includes the interface with the communication channel but does not include the interconnecting channel.

MODEM: An acronym for modulator/demodulator. A hardware device used for changing digital information to and from an analog form to allow transmission over voice grade circuits.

MONITORING (SECURITY MONITORING): Security monitoring is the on-line identification and the dynamics display of the actual operating conditions of the power system.

MULTIPLEX: To interleave or simultaneously transmit two or more messages on a single channel.

NETWORK TOPOLOGY: The relative position of the ideal elements representing an electric network.

OBSERVABILITY: A power system is said to be observable, in the static state estimation sense for a given measurement set M, if the bus voltage magnitudes and angles throughout the system can be determined by processing the measurements in M by a static state estimator. Determination of the observable parts of the power system, based on the results of topology and available analog measurements.

OPEN-LOOP: A control system in which the system outputs are controlled by system inputs only, and no account is taken of actual system output.

OPTIMAL POWER FLOW: Determination of corrective actions to achieve minimal active power flow considering security constraints.

PARALLEL PROCESSING: Simultaneous processing of various commands or rules, as opposed to the sequential processing in conventional (von Neumann) type of computer architecture (see von Neumann architecture).

PATTERN RECOGNITION: A principle of computerised problem solving that aims at classifying data into predetermined categories.

PERFORMANCE REQUIREMENT: A requirement that specifies a performance characteristic that a system or system component must possess; for example, speed, accuracy, frequency.

PLAUSIBILITY: Plausibility check of breaker position and analog measurements.

POWER APPLICATION SOFTWARE (PAS): Set of functions aiding the operator in his decision-making process. The PAS domain is crossing the traditional boundaries between: power system engineering; automatic control engineering; numerical analysis; software engineering. PAS is a multidisciplinary domain or field of knowledge.

POWER SYSTEM: A power system can be considered as a complex process in which energy is transformed from a fossil, nuclear, or hydro state to electricity, is transmitted and distributed and finally delivered to be transformed into useful work.

POWER SYSTEM PLANNING: The whole range of studies involved in the development of a technically and economically sound system of electricity supply.

POWER SYSTEM STABILITY: The capability of a power system to regain a steady state, characterised by the synchronous operation of the generators after a disruption due, for example, to variation of power or impedance.

PROTECTIVE RELAYING: The process of implementing equipment, or both, for remote utilisation of the control circuit of a detector of defective lines or apparatus or other condition of an abnormal or dangerous nature.

PROTOCOL: The method or code describing adherence to the order, preference, and procedure used in a control system. It describes the method structure used to communicate the elements of the control system.

RADIAL OPERATION: The method of operation where each point of a given part of a network is fed along one path only.

REACTIVE POWER COMPENSATION: An action to optimise the transmission of reactive power in the network as a whole.

REACTIVE-POWER VOLTAGE CONTROL: Voltage control by the adjustment of reactive power generation in a power system.

REAL-TIME: Pertaining to the processing of data by a computer in connection with another process outside the computer according to time requirements imposed by the outside process. This term is also used to describe systems operating in conversational mode, and processes that can be influenced by human intervention while they are in progress.

REDUNDANCY: The existence of two or more means for carrying out a specified function.

REDUNDANCY FACTOR: Numerical value characterising the set of measurements available for the state estimation of a network state variable 2n-1; n denotes the number of buses: $\eta = m/(2n-1)$.

REFERENCE NODE: The node of a network where the phase of the node voltage in the complex plane is arbitrarily fixed, in reference to which the phases of the system state variables are measured.

REGULATING ENERGY OF A SYSTEM; POWER/FREQUENCY CHARACTERISTIC: For a power system the quotient of the change of active power demand by the corresponding change of frequency in the absence of secondary control.

RELATIONAL MODEL: A means of describing the data structure in which all data is shown as belonging to simple tables. No access paths are predefined and relationships exist only by virtue of certain fields in one table having the same value as certain other fields in another table.

RELIABILITY: The probability that an item will carry out its required mission satisfactorily for a specified period off time when used according to designed conditions.

REMOTE TERMINAL UNIT (RTU): Refers to the remote station equipment of a supervisory system.

RIPPLE CONTROL: A low frequency (100-500 Hz) communications system installed within an electric power network to switch load.

SCADA-SUPERVISORY CONTROL AND DATA ACQUISITION: Twenty years were required before the term SCADA gained acceptance and some measure of definition. The generally accepted idea of the SCADA system is a mixture of some of the following: several processors at remote sites that transmit data either on demand or on the occurrence of an event (Remote Terminal Units); a central supervisor to provide higher level load computation, generally a master or a host computer; a long distance transmission system between the remote terminal units (RTU) and the central supervisor; a coded system to assure the security and reliability of the transmitted information. The SCADA is based on the "one-sensor-one-signal" philosophy. The analysis is based on primary process data consisting of binary status indications, limit excess information or analog values.

SECONDARY CONTROL: Coordinated control of the active power supplied to the network by particular generators.

SECURITY ASSESSMENT: The security of the power system operation is assessed by comparing the operation of the individual elements and the system as a whole against predefined limits.

SECURITY MONITORING: To calculate a complete, reliable and consistent set of data, taking into consideration the influence of neighbouring systems, to provide accurate information about the actual situation in the power system, independent of measurement errors and measurement outages.

SHORT CIRCUIT CALCULATION - ASYMMETRICAL: Calculation of asymmetrical short circuit power and checking against breaking capacity of the circuit breakers.

SHORT CIRCUIT CALCULATION - SYMMETRICAL: Calculation of symmetrical short circuit power and checking against breaking capacity of the circuit breakers.

SHORT-CIRCUIT CALCULATION: Computation of the currents and voltages within a network subjected to a short circuit.

SHORT-CIRCUIT CURRENT: The current flowing at a given point of a network resulting form a short-circuit at another point of this network.

SHORT-CIRCUIT CURRENT STABILITY: The permissible value of the short-circuit current in a given network component for a specified duration.

SIMPLEX CIRCUIT: A communication channel capable of transmitting data signals in one direction only.

SIMULATION: The representation of selected characteristics concerning the behaviour of one physical or abstract system by another system. In a digital computer system, simulation is done by software.

SIMULATOR: A device, data processing system, or computer program that represents certain features of the behaviour of a physical or abstract system.

SLACK BUS: An infinite bus where the voltage magnitude is predetermined and which is at the same time a reference node and balancing bus.

STABILITY: Transient stability assessment consists of determining whether the system's oscillations following an outage or a short circuit fault will cause loss of synchronism.

STATE DETERMINATION: Is designed to determine the current operating state of the power system based on real-time measurements. The aim of the static state estimation (SE) is to calculate unmeasured quantities, to improve measured quantities and to detect bad measurements and hence to produce a complete, consistent and reliable set of data, based on the total redundant measurements. The state estimation function determines the "best" estimate of the network state (complex node voltages which minimise the sum of the weighted squared measurement error). SE is a digital processing scheme which provides a real-time data base for many of the central control and dispatch decisions necessary in operating a power system. SE may be defined as the process of computing the network node voltage and then line flows, node loads, etc. from measurements made upon the network at a sampling time. SE transforms the meter readings into an estimate of the system state or static state vector. SE is an algorithm that estimates the state of the system from acquired measurements and known physical characteristics.

STATE ESTIMATION: The computation of the most probable currents and voltages within the network at a given instant by solving a system of mostly non-linear equations whose parameters are obtained by means of redundant measurements.

STATUS: Information describing a logical state of a point or equipment.

STEADY STATE OF A SYSTEM: The operating conditions of a network in which the system state variables are considered to be sensibly constant.

STEADY STATE STABILITY OF A POWER SYSTEM: A power system stability in which any disruptions occurring only have small rates of change and small relative magnitudes.

SUPERVISORY CONTROL: An arrangement for selective control and supervision of remotely located apparatus using multiplexing techniques.

SUPERVISORY CONTROL AND DATA ACQUISITION (SCADA): A set of functions that acquires and processes real-time system data and presents them to the system operator. The operator also uses this set of functions to remotely control equipment in the power system.

SUPERVISORY CONTROL FUNCTIONS: Equipments governed by this standard comprise one or more of the following functions: alarm function; analog function; control function; indication (status) function; accumulator function; sequence of events function.

SYSTEM PARAMETERS; SYSTEM CONSTANTS: The quantities considered as being fixed which characterise the system components. Examples: impedances, admittances, transformation ratio.

SYSTEM STATE VARIABLES: The variable quantities associated with the electrical state of the system. In general, the system state variables are identical to the complex bus voltages. With system state variables all other electric quantities (current power) can be calculated.

TIME SERIES: A set of numerical values that represents the level and the status of some ongoing activities over time.

TOP-DOWN: Pertaining to an approach that starts with the highest level component of a hierarchy and proceeds progressively through lower levels.

TOPOLOGICAL DIAGRAM OF A NETWORK: The pictorial representation of the network topology.

TOPOLOGY: The topology function determines the connectivity of the network based on the specified positions of the switches. The resultant node/branch model is used by a number of other PAS functions. The topology function is designed to determine the connectivity of the power system network based on the position of the switches. The topology function is needed to support all of the network analysis programs; they require that the network structure be defined in terms of nodes and branches. Using the element model, the topological connections and the actual switch status, the topology program builds a node/branch model.

TRANSIENT STABILITY OF A POWER SYSTEM: A power system stability in which any disruption may have large rates of change and/or large relative magnitudes.

TRANSIENT STATE OF A SYSTEM: The operating condition of a network in which at least one of the state variables is changing, usually for a short period.

TRANSMISSION LOSS CALCULATION: Calculation of transmission loss coefficients for economic dispatch calculation, based on the actual network condition.

UNIT COMMITMENT: Planning of the operation of thermal power stations, based on load forecasting, start-up costs, operation costs, stand-still costs, etc.

VIRTUAL MEMORY: Storage space that is the peripheral or auxiliary memory space addressable by the user; thus virtual memory is perceived to be main storage space. A programming method that allows the operating system to provide essentially unlimited program address space. In a VAX-11 computer, the virtual memory design means that a VAX-11 program can address more than 4 gigabytes of address space.

VIRTUAL MEMORY/VIRTUAL STORAGE: The storage space that may be regarded as the addressable main storage by the user of a computer system in which virtual addresses are mapped into real addresses.

VOLTAGE REDUCTION: As a combined or coordinated function of substation and feeder automation, the process of, or equipment for temporary conservation of energy, implemented by reducing nominal electric distribution voltage by a selected percentage.

VOLTAGE/REACTIVE POWER OPTIMISATION: The determination of corrective actions to achieve the minimal reactive power flow and to observe the proper voltage profile taking into consideration security constraints.

VON NEUMANN ARCHITECTURE: Most of the computers nowadays are of an architecture that uses sequential processing. Somehow simplifying the lines within computer program source code might be understood as a sequence of commands executed by the machine one after another (see parallel processing).

WHEELING: A technique by which a utility, located between two others, transfers power from its neighbour on one side to its neighbour on the other.

WHEELING SERVICE: The use of and access to the transmission facilities of one system to transmit power or energy of and for another non-continuous system or user.

ZOOM: In computer graphics, continuously scaling the display elements of display image to perceive and manipulate details more clearly which are not readily perceived in the previous view.

References

Selected papers

Scheidt, J.L.: "A Survey of Power System Control Center Justifications", IEEE, Vol.PAS-98, No 1, Jan/Feb 1979

IEEE: "Definitions, Specifications and Analysis of Manual, Automatic and Supervisory Station Control and Data Acquisition", ANSI/IEEE c 37.1-1979

Conradie, J.F.D.: "Experience gained during the assembly phase of ESCOM's System Control and Load Dispatch System", CIGRE-Symposium, Study Committee 32, Minneapolis, 1979

Dy Liacco, T.E.: "New Developments in Computer Configurations for Control Centers", Power System Monitoring and Control, IEE Conf. Publ. No 187, London, June 24/26, 1980

Wollenberg, B.F. et al.: "Cost-Benefit Justification of an Energy Control Center", IEEE Summer Power Meeting 534-8, 1980

Mäntynen, R. and Värsänen, J.: "Development of Power System Control in Imatran Voima", paper 32-06, CIGRE Session, Paris, 1980

Maier, D.W. et al.: "Standardization SCADA-AGC System. Effect on System Cost, Delivery and Quality", IEEE, Vol. PAS-101, No 1, Jan 1982

EPRI: "Human Factors Review of Electric Power Dispatch Control Centers", EPRI Report, EL-1960, Res. Project No. 1354-1, Oct 1982

EPRI: "Software Development and Maintenance Guidelines", EPRI-Report, EL-3089, Res. Project No 1714-1, May 1983

Duncan, G.L. and Gorsha, R.A.: "Project Management. A Major Factor in Project Success", IEEE, Vol PAS-102, No 11, Nov 1983

EPRI: "Guidelines for Evaluating Distribution Automation", EPRI Report, EL-3728, Res. Project No 2021, 1984

Grein, W.; Müller, H. and Petroianu, A.: "Dynamic Network Colouring", Proc. of PSCC, Helsinki, Aug 19 - 24, 1984

Imhof, K.: "Graph Theory Approach to Power System Modelling and Analysis", Proc. of PSCC, Helsinki, 1984

EPRI: "Computer-Generated Display System Guidelines", EPRI Report, No 37-1, Res. Project No 3184, Sep 1984

Masiello, R.D.: "Computers in Power, A Welcome Invador", IEEE Spectrum, Feb 1985

Aboytes, F. and Arroyo, G.: "Security Assessment in the Operation of Longitudinal Power Systems", Proc. of PICA, San Francisco, May 6 - 10, 1985

van Nielen, N.S.: "Report on Data Exchanges Between Control Centres", UNIPEDE-Congress, Athens, June 9 - 14, 1985

Barret, J.P. and Waha, J.P.: "Future Automation in Transmission Networks", UNIPEDE-Congress, Athens, June 9 - 14, 1985

Daniels, H. et al.: "More than mainframes", IEEE Spectrum, August 1985

CIGRE: "Electric Power Systems in Developing Countries", Proc. of CIGRE-Symposium, Dakar, Nov 1985

Mariani, E. and Wiklund, L.: "Hierarchical Organisation of Control Centres", CIGRE - Working Group 02, ELECTRA, Vol. 98, No 1, 1985

Mc Dyer, F. and Herger, K.: "The New National Control Centre, Ireland", IEEE Vol. PWRS-2, No 1, Feb 1987

Schaffer, G.: "User-oriented Software in Energy Management Systems", Proc. of PSCC, Cascais, Portugal, Aug 31 - Sep 4, 1987

Singer, M.: "The Application of a Database Management System in an Energy Management System", Proc. of PSCC, Cascais, Portugal, 1987

Amelink, H. et al.: "Data Flows, Data Processing and Data Presentation for Power System Operation and Control", CIGRE-Symposium, Study Committee 39, Tokyo, Oct 26-31, 1987

Schwab, B.C. et al.: "Software Features of a DBMS Kernel in Large Real-Time EMS Systems", IEEE, Winter Power Meeting, WM 089-6, 1987

Hanson, V.: "Some Aspects of Computer Loading Problems in Modern Control Centres", ELECTRA, No 114, 1987

Dy Liacco, T.E.:"Survey of System Control Centers for Generation-Transmission Systems", The Dy Liacco Corporation, 1987

Baer, W.: "Technology and the Business of Information", EPRI-Journal, April/May 1988

Handschin, E. and Knight, U.G.: "System Operation Training Simulators", CIGRE Session, Paris, 1988

Thurein, I.P.: "Adequacy of System Security Assessment Tools for System Operation", paper 39-02, CIGRE Session, Paris, 1988

CIGRE: "Operation of Electric Power Systems in Developing Countries", Proc. of CIGRE-Symposium, Bangkok, Nov 1989

Selected books

Anderson, P.M.: "Analysis of Faulted Power Systems", Iowa State University Press, 1973

Berrie, T.W.: "Power System Economies", Peter Peregrinus Ltd., London, 1983

Blackburn, J.L.: "Protective Relaying", Marcel Dekker, Inc., 1987

Bohn, Th. and Bitterlich, W. (Ed.): "Handbuchreihe Energie (17 Bände)", Technischer Verlag Resch, Verlag TÜV Rheinland, 1982

Cegrell, T.: "Power System Control Technology", Prentice Hall, 1986

Debs, A.S.: "Modern Power Systems Control and Operation", Kluwer Academic Publishers, 1988

Elgerd, O.I.: "Electric Energy Systems Theory: An Introduction", McGraw Hill, 1972

Fouad, A. and Anderson, P.M.: "Power System Control and Stability", Iowa State University Press, 1977

Glover, J.D. and Sarma, M.: "Power System Analysis and Design", PWS Publishers, Boston, 1987

Gönen, T.: "Modern Power Systems Analysis", Wiley Interscience, 1987

Gross, C.A.: "Power System Analysis", Wiley, New York, 1979

Happoldt, H. and Oeding, D.: "Elektrische Kraftwerke und Netze", Springer Verlag, 1978

Handschin, E. (Ed.): "Real-Time Control of Electric Power Systems", Elsevier Publishing Company, 1972

Handschin, E.: "Elektrische Energieübertragungssysteme", Dr. Alfred Hüthig Verlag, 1987

Hosemann, G. and Boeck, W.: "Grundlagen der elektrischen Energietechnik", Springer Verlag, 1979

Hütte, Taschenbücher der Technik: "Elektrische Energietechnik", Band 3, Netze, Springer Verlag, 1988

Kleemann, M. and Meliß, M.: "Regenerative Energiequellen", Springer Verlag, 1988

Knight, U.G.: "Power Systems Engineering and Mathematics", Pergamon Press, 1972

Minakewa, I.: "Integrated Analysis System for Power System Planning and Operation", in CAD/CAM and MIS IN JAPAN, Ed. Noboru Takagi, John Branch, Academic Press, 1986

Rumpel, D. and Ji Rong Sun: "Netzleittechnik", Springer Verlag, 1989

Russel, D. (Ed.): "Power System Control and Protection", Academic Press, 1978

Sakis Meliopoulos, A.P.: "Power System Grounding and Transients", M. Dekker, Inc., 1988

Schäfer, H.: "Elektrische Kraftwerkstechnik", Springer Verlag, 1979

Stevenson, W.D.: "Elements of Power System Analysis", McGraw Hill, 1975

Weedy, B.M.: "Electric Power Systems", Wiley, 1972

Wood, A.J. and Wollenberg, B.F.: "Power Generation, Operation and Control", Wiley, New York, 1984

MIX
Papier aus verantwortungsvollen Quellen
Paper from responsible sources
FSC® C105338

If you have any concerns about our products,
you can contact us on
ProductSafety@springernature.com

In case Publisher is established outside the EU,
the EU authorized representative is:
Springer Nature Customer Service Center GmbH
Europaplatz 3, 69115 Heidelberg, Germany

Printed by Libri Plureos GmbH
in Hamburg, Germany